Smart Innovation, Systems and Technologies

Volume 300

Series Editors

Robert J. Howlett, Bournemouth University and KES International, Shoreham-by-Sea, UK

Lakhmi C. Jain, KES International, Shoreham-by-Sea, UK

The Smart Innovation, Systems and Technologies book series encompasses the topics of knowledge, intelligence, innovation and sustainability. The aim of the series is to make available a platform for the publication of books on all aspects of single and multi-disciplinary research on these themes in order to make the latest results available in a readily-accessible form. Volumes on interdisciplinary research combining two or more of these areas is particularly sought.

The series covers systems and paradigms that employ knowledge and intelligence in a broad sense. Its scope is systems having embedded knowledge and intelligence, which may be applied to the solution of world problems in industry, the environment and the community. It also focusses on the knowledge-transfer methodologies and innovation strategies employed to make this happen effectively. The combination of intelligent systems tools and a broad range of applications introduces a need for a synergy of disciplines from science, technology, business and the humanities. The series will include conference proceedings, edited collections, monographs, handbooks, reference books, and other relevant types of book in areas of science and technology where smart systems and technologies can offer innovative solutions.

High quality content is an essential feature for all book proposals accepted for the series. It is expected that editors of all accepted volumes will ensure that contributions are subjected to an appropriate level of reviewing process and adhere to KES quality principles.

Indexed by SCOPUS, EI Compendex, INSPEC, WTI Frankfurt eG, zbMATH, Japanese Science and Technology Agency (JST), SCImago, DBLP.

All books published in the series are submitted for consideration in Web of Science.

Sergey V. Zykov

IT Crisisology Casebook

Smart Digitalization for Sustainable
Development

 Springer

Sergey V. Zykov
National Research University Higher
School of Economics
Moscow, Russia

ISSN 2190-3018   ISSN 2190-3026 (electronic)
Smart Innovation, Systems and Technologies
ISBN 978-981-19-2233-6   ISBN 978-981-19-2231-2 (eBook)
https://doi.org/10.1007/978-981-19-2231-2

This Springer imprint is published by the registered company Springer Nature Singapore Pte Ltd.
The registered company address is: 152 Beach Road, #21-01/04 Gateway East, Singapore 189721,
Singapore

To God, my teachers, and my family

Foreword

As a professor, I am always looking for books that offer a new perspective on the content that I teach. Because most of my students are executives, my attention is particularly drawn towards books that present materials in a manner that relates directly to the world as they experience it, sometimes referred to as the "real world". For this reason, I was delighted to be afforded the opportunity to write the Foreword to Sergey Zykov's newest book, "IT Crisisology Casebook: Smart Digitalization for Sustainable Development".

The central theme of this book is "IT Crisisology". The underlying idea is a simple one. As organizations, and the environments with which they interact, grow more digital, they must transform to survive. Unfortunately, such transformations are nearly always precipitated by, or followed by, a crisis. Making it through a crisis is challenging—the process can lead to either value creation or value destruction. How the organization responds will determine which.

To improve our odds of thriving through transformation-wrought crises, it makes sense to learn about them. Unfortunately, general rules for dealing with IT transformation crises are few and far between. To paraphrase the famous opening line of Anna Karenina: "Organizations in the steady state are all alike; every organization in crisis experiences that crisis in its own way". A regrettable truth in a complex world.

So, if we cannot teach general rules for managing the crises of IT transformation, how do we learn about them? One way is by reasoning through examples; by viewing complex situations through the eyes of the managers that confronted them. That is where case studies come in. Rather than providing the reader with rules, these cases offer readers the opportunity to think through a situation and come up with their own situation-specific rules.

The benefits of using case studies do not end with helping readers hone their judgement. Storytelling is the most resonant means of informing—one that works effectively for everyone from the youngest child to the most seasoned executive. Choose the right stories and you not only inform the reader but also inform those to whom the reader retells the stories. And the collection of stories contained in this book should resonate with nearly everyone interested in digitalization and how it is impacting organizations and the environment. No wonder I found it exciting to read.

The book is organized around evolving stages of digitalization and the many challenges the process presents. After setting out a framework and demonstrating how case studies can be applied in the initial chapter, the second chapter begins by looking at early stage transformations, focusing on a series of companies many of which are household names: the consulting company Accenture, the entertainment companies Cirque du Soleil and Disney, the technology companies Dropbox and Foursquare, and the fashion retailer Zara. Understanding how these organizations evolved and transformed will improve both the readers' understanding of the process and their overall IT literacy.

The next chapter begins by focusing on some small businesses that have used digitalization in innovative ways to support their strategies. It then considers some well-known and lesser-known examples of organizations that experienced transformation crises and did not handle them effectively, including Eastman Kodak, ThyssenKrupp, and Blockbuster Video, contrasting these with organizations that managed to pull away from the brink of failure by reinventing themselves for a digital world, such as LEGO.

The fourth chapter, which focuses on multinational organizations and diversity, presents the story of Microsoft, which initially neglected the digitalization process but managed to catch up through new leadership. It also tells the story of Huawei, whose adherence to a core set of principles helped it avoid the declines experienced by so many of its competitors in the mobile device industry, including Deutsche Telecom, Alcatel, Lucent, and Motorola.

In Chap. 5, Zykov turns his focus to an industry case focusing on the Russian forest industry. The case illustrates how issues of technology, resources, the environment, businesses, and government all play an integral part in the transformation process and the crises it produces. The chapter also looks at the CMM(I) model of software process evolution, a transformation-driven model that reflects the degree of control a software development shop exerts over its activities.

The sixth chapter looks at the human side of digitalization, considering the question of how knowledge is transferred and the potential need to accommodate cultural differences. It also looks specifically at the soft skills needed by a software developer.

In the conclusion, the lessons of these chapters are pulled together and some general thoughts about dealing with complex situations are presented.

Upon reading this book, I gained a great appreciation for the many forms in which transformation can manifest itself and in which the accompanying crises can be addressed. I am sure that the reader will feel the same.

T. Grandon Gill, DBA
Professor and DBA Academic Director
School of Information Systems
and Management
Muma College of Business
University of South Florida
Tampa, FL, USA

Preface

The focus of this book is real-world case-based crisis management in digital product development. This includes forecasting, responding, and agile engineering/management methods, patterns, and practices for sustainable development.

Over the past decades, production in general and digital product development in particular were understood and practised in different ways. Changeable business constraints, complex technical requirements, and the so-called "human factors" imposed on the digital products caused what was articulated as sustainability "crises". These complex sources of trouble require a practical multifaceted approach to address each of their ingredients. Therefore, this book suggests an approach that contains practical methods, patterns, and techniques to efficiently manage these crises and provide sustainable development.

Software engineering was triggered by what was initially identified as a digital production "crisis"; however, this practically focused discipline even after 50 years of existence cannot be considered a "silver bullet" for digitalization, and wider, sustainable organization development. This means that the digital development/production crisis is still here, and it may immediately occur in case of careless selection or unbalanced application of the rich variety of the principles and practices that the state-of-the-art digital product engineering currently incorporates.

This book introduces a set of case studies for sustainability in management as a blend, the components of which have been carefully selected from a few domains adjacent to digital production such as IT-intensive operation, human resource management, and knowledge engineering, to name a few. The key ingredients of this crisis management framework include information management, tradeoff optimization, agile product development, and knowledge transfer.

The case studies this book features, will help the stakeholders in understanding and identifying the key technology, business, and human factors that may likely result in a digital production crisis, i.e., critically affect the organization outcomes in terms of successful digitalization and sustainable development. These factors are particularly important for large-scale applications, typically considered very complex in managerial and technological aspects, and therefore specifically addressed by the discipline of IT Crisisology. Therefore, this book will throw light on the crisis responsive and resilient methods, techniques, and practices; as such, it will focus on their practical and realistic applications and the resulting benefits for digitalization and sustainable development.

To successfully apply the social and human aspects of IT Crisisology, which often appear subtle, uncertain, and hardly manageable, this book suggests the case study-based approach. An extensive set of comparative case studies for IT-intensive digital businesses of different scales and scopes will be considered. We approach the crisis management solutions from the perspectives of different continents, historical and cultural diversity, which can essentially affect the human factors that often are the root cause of a crisis. The businesses that we examine in these case studies clearly have a number of similarities including their overall structure and ultimate goals. However, certain outcomes and business deliverables due to local varieties and business-specific dynamics might be essentially different. After discussing each of these case studies separately (from the perspectives of business processes, knowledge transfer, and digital products utilized), we will compare them in terms of business, technology, and human-related factors to detect and refine common patterns of digitalization sustainability in crisis environments.

We hope that this book will serve as a reliable compass for the digital product developers and managers of IT-intensive businesses as it will give them the necessary guidelines to navigate confidently through the rough ocean of digitalization in the stormy times of crises.

Moscow, Russia   Sergey V. Zykov

Acknowledgements

I would like to thank my colleagues who significantly contributed to this book. They clarified my initially vague concepts and assisted in a number of processes including translation, copyediting, diagramming, etc. These are the students who did their master's/Ph.D. theses under my supervision. A few of their takeaways were transformed and included in this book as case studies on agility improvement. They are Joseph Attakorah, Yaw Buadu, Maxim Gilman, Andrey Ivanov, Prince Islam, Yao Jianlong, Daria Krasnova, Alexander Lazarenko, Nikita Morgun, Evgenia Murzaeva, Vassily Naumushkin, Vadim Piven, Nikita Rubinov, Nikita Shaimov, Valeriia Shevchenko, Indra Singh, Daria Vikulova, Nikita Zaytsev, Alexey Zheleznoy, and Maria Zolotukhina.

I would like to thank the Springer Editorial Director Dr. Thomas Ditzinger, the Springer Executive Editor Mr. Aninda Bose, the Springer Senior Executive for Production Mr. Ashok Kumar, and the Springer Project Coordinators for Books Production, Mr. Daniel Joseph Glarance and Mr. Gowrishankar Ayyasamy, for their continuous availability and prompt assistance.

In addition, I would like to express my deep appreciation and sincere gratitude to the Editors-in-Chief of the Springer Series in Smart Innovation, Systems and Technologies, Prof. Lakhmi C. Jain and Prof. Robert J. Howlett, for their tireless efforts in supporting my ideas.

Contents

About the Author

Prof. Dr. Sergey V. Zykov holds a Ph.D. (2000) and Dr. Habil. (2017) in Computer Science. He has a 20-year experience in IT, including Vice-CIO of the ITERA International Oil & Gas Group. He also has over 20 years in teaching computer science and software engineering and holds instructor certificates from Carnegie Mellon University and London School of Economics. Currently, he is a Full Professor at the HSE University, National Nuclear Research University MEPhI, and Russian Technical University MIREA. He served as a Visiting Researcher at the Carnegie Mellon University (USA), the First Moscow State Medical University, and Innopolis University (Russia). He authored over 100 papers and 10 books, including 5 monographs by Springer, among which is IT Crisisology: Smart Crisis Management in Software Engineering (2020). He serves as an Associate Editor at the Intelligent Decision Technologies and International Journal of Knowledge-Based and Intelligent Engineering Systems. His research fields include: crisis responsive software development, enterprise system lifecycles, and data modeling.

Acronyms

6σ	*Six Sigma*
ACDM	Architecture-Centric Development Method
AHP	Analytic Hierarchy Process
API	Application Programming Interface
AR	Augmented Reality
BCG	Boston Consulting Group
BI	Business Intelligence
BiTA	Blockchain in Transport Alliance
CBA	Choosing by Advantages
CIO	Chief information officer
CMM	Capability Maturity Model
CMMI	Capability Maturity Model Integration
CRM	Customer Relationship Management
CSF	Critical Success Factor
DEA	Data Envelopment Analysis
DMAIC	Define, Measure, Analyze, Improve, and Control
DMU	Decision-Making Units
EAM	Enterprise Agility Matrix
EBIT	Earnings Before Interest and Taxes
ERA	Evidential Reasoning Approach
ERP	Enterprise Resource Planning
FIIF	Forest Industry Innovation Framework
GDP	Gross domestic product
GUI	Graphical User Interface
ICT	Information and communication technologies
IDC	International Data Corporation
IofAs	Importance of Advantages
IoT	Internet of Things
IQR	Interquartile Range
ISO	International Organization of Standardization
ITC(F)	IT Crisisology (Framework)

KM	Knowledge Management
KPI	Key Performance Indicator
KS	Knowledge Sharing
KT	Knowledge Transfer
KTT	KM Tools and Technology
M2M	Machine-to-Machine
MCDA	Multiple-Criteria Decision Analysis
MCS	Mobile Crowdsensing
MNC	Multinational Corporation
NTI	National Technology Initiative
NYU	New York University
OS	Operating System
PDCA	Plan-Do-Check-Adjust
PDM	Precedence Diagramming Method
PPE	Personal Protective Equipment
PSM	Process Safety Management
PSP	Personal Software Process
PtD	Prevention Through Design
QA	Quality Attribute
RFID	Radio Frequency Identification Technology
ROI	Return On Investment
SCADA	Supervisory Control And Data Acquisition
SD	System Dynamics
SDK	Software Development Kit
SDLC	Software Development Lifecycle
SEI	Software Engineering Institute
SME	Small and Medium Enterprises
STM	Science, Technology, Medicine
TMCS	Top Management Commitment and Support
TNC	Transnational Corporation
TOGAF	The Open Group Architecture Framework
TSP	Team Software Process
UI	User interface
UX	User experience
VR	Virtual Reality
XR	Extended Reality

Chapter 1
Introduction: The Crisis of Digitalization

Digitalization essentially involves a number of aspects to be addressed concurrently in order to avoid transformation crises. These key aspects are process, data, and IT system transformation. For each of the aspects, these changes involve a number of layers in large-scale and therefore complex enterprises. To address these multi-faceted transformation processes, we suggest using the Enterprise Agility Matrix (EAM) introduced in our previous books [5, 6]. In view of digitalization, these matrix values are being transformed so that the resulting enterprise architecture (in terms of components and connectors) becomes more agile and therefore crisis-resistant. Consequently, organization maturity improves. This maturity level increase is typically modelled by the CMM/CMMI framework [2]. Thus, the CMMI-level upgrade (e.g., from ad hoc to be quantified and further to be optimized, etc.) can be assigned in terms of EAM change. Therefore, it is possible to represent each transformation step in terms of EAM changes, which include certain improvements in data, process, and system design management.

Visually, this can be represented by a 3D model, one axis of which includes maturity levels, and the rest two correspond to the EAM dimensions of component-data-process, and infrastructure levels (such as hardware, SCADA, EAM, ERP, BI, etc.). The organizational progress in maturity can be traced by analysing connector dependencies of the EAM before and after each transformation step. This can be further improved by detecting the "pain points" in terms of dependencies and adjusting these to make the infrastructure more crisis responsive and resistant.

The above general framework involves a number of related aspects, such as business process optimization, system architecture management, and knowledge engineering, to name a few.

The ITC framework [7] systematically addresses these multiple aspects by means of a family of carefully selected approaches such as

- ACDM (architecture-centric development method),
- KM/KT (knowledge management/transfer),
- PAEI (enterprise/organization lifecycle model),
- Informing science models (based on Shannon information theory),

S. V. Zykov, *IT Crisisology Casebook*, Smart Innovation, Systems and Technologies 300, https://doi.org/10.1007/978-981-19-2231-2_1

- "7 Principles" and other approaches to knowledge transfer,
- Enhanced Spiral model (see also [5, 6]).

Further enhancement of the ITC in terms of organizational and business management should involve additional approaches such as The Open Group Architecture Framework (TOGAF), etc.

An essential part of this ITC approach is case studies, the key aspects of which include:

- Creating case study (based on student/researcher experience and/or original data).
- Composing questions to discuss the case study.
- Producing a follow-up ("postmortem") upon discussing case studies.
- Formally modeling case studies (possible approaches include concept-relationship maps, process/data graph-based notations, etc.).
- Developing software/tools for semi-automated case study generation (such as concept maps, questions/topics for discussion, walkthrough, main text, etc.).

The key aspects of case study application to problem-solving in general and decision-making in particular are summarized in [3]. Often, this process includes multiple criteria optimizations, and tradeoff-based methods, which are central for crisis management in general and software engineering in particular.

We address these methods in our previous books, particularly in [7]; examples include:

- analytic hierarchy process (AHP),
- choosing by advantages (CBA),
- data envelopment analysis (DEA),
- evidential reasoning approach (ERA),
- architecture-centric development method (ACDM).

Also, IT-intensive crisis management essentially involves harnessing human-related factors and using them in a "resonant" manner in order to ensure well-informed decisions [7]. Case studies as a powerful tool provide a great value to master human factors and their wise application to crisis management by means of ITC framework. Although we introduced a few examples of case study assistance in crisis management of IT-intensive businesses, for systematic application, these clearly need a more substantial coverage being the core subject of a book. Besides, a clearer focus is required for digital transformation to ensure sustainable development.

One powerful approach that studies the transformations in a formal way is the lambda calculus. This approach was suggested by Church (and originally published in the book titled "The calculi of lambda conversion" in Annals of Mathematics Studies) in 1941 [1]. This calculus is based on the earlier results outlined by Schönfinkel at the beginning of the last century and is therefore around 100 years old [4]. In the following decades, the approach was enhanced for computational applications (though computers did not exist, to say nothing about networking, the internet, and digitalization), and became Turing complete. The key idea is that a mathematical function can represent an object as a model and being applied to another function can represent a complex object. Also, we can refer to any function by its signature (i.e., name and parameters); this is called abstraction operator. Clearly, these two operations transform the initial objects that they accept as arguments (i.e., inputs). Irrespective of the initial object complexity, we can represent it conveniently by means of abstraction and instantiate it by means of the application operator. Moreover, to represent any object, we need only two primitive objects. One is used to split an object into its head and tail, the other aggregates two objects in a chain and tags each of them by the third object. The first transformation is somewhat similar to clipping headline of a news item. The second transformation resembles joining two members in a community and marking (tagging/flagging) each by the community label. Therefore, this "divide-and-conquer" approach assists in complexity management, which is essential for crisis-resistant development both in terms of software engineering and organizational management.

In this view, for an organization (or software product) of virtually any complexity, there are a few ways to transform and represent it (e.g., detailed/styled differently). The Bhagavad Gita presents Krishna as a creator-redeemer god who creates a real world of souls and matter. In this context, the Shiva god is destroyer and Vishnu is guardian-preserver. Ironically, the combinators (i.e., lambda calculus objects that contain no free variables) typically named K and S act oppositely to the above gods' initials: K is the "destroyer" and S is the "creator". These two make the minimal basis, i.e., the combinator set allowing the construction of any combinator object (i.e., an object of arbitrary complexity). However, this "two-word language" often overcomplicates "sentences" and often generates bulky objects containing complexly structured combinations of S and K in deeply nested brackets. A more common basis also includes the I (i.e., identity), which can be called the Vishnu of this combinator "world" as it keeps many of these bulky objects simpler. In terms of "Shiva" and "Krishna", the preserving "Vishnu" combinator is represented as follows: I = S(KS)K.

A critical question arises: how many items should the formal notation include in order to keep complicated objects efficiently represented in terms of simplicity, intuitiveness, visibility, comprehensibility, etc.? More elaborate bases include operators for permutation, duplication, etc. To model multiple stage transformations, approaches that are more elaborate also include iteration (often represented mathematically as recursion).

The formal models mentioned are applicable not only to transformations but also for case study representation in terms of key concepts and relationships, including evolving events and object attributes. These are graphically represented by diagrams such as mind maps and frames.

Clearly, this transformation-based approach choice essentially depends on the project scale and scope and is a problem of multiple criteria optimization that requires tradeoffs. Software engineering assists in this tradeoff-based IT-intensive digital product development; we have addressed its crisis-related issues by the ITC framework presented in [8].

Coming back to the case studies, our approach is largely inspired by Gill's approach to the "case method" as he calls it in his monograph [3]. Gill introduces a few mission-critical principles and practices of building case studies and "informing" both students and experts (including researchers and practitioners) based on these. He is a famous US Professor in Information Systems and Business Administration (also heading DBA Program at the University of South Florida) who had a rich and extensive multiple context experience with case studies as a research and practice method.

When asked about a case study meaning, we often refer to such genres as stories, parables, myths, legends, fables, or even fairy tales. In fact, Gill also used a tale of the three little pigs (that we will also refer to further) to illustrate the approach and its applications (i.e., use cases) in his casebook. The idea of a case study is that it clearly describes a realistic or even a real-world situation that contains open issue(s)/question(s). However, it never has a single right solution; instead, at least a few feasible and applicable solutions are derivable from this story. This is due to a tradeoff nature of each solution (each of which requires a rather complex multi-criteria analysis) and therefore is not a strictly optimal one, but rather a "good enough" option, as they call it at Carnegie Mellon University, the birthplace of software engineering.

This book is structured as follows. Chapter 1 introduces the concept of crisis and other essential aspects.

Chapter 2 elaborates on digitalization in view of case-based reasoning; it surveys digitalization approaches, principles, and practices in view of the IT Crisisology (ITC) framework introduced in our previous books. Our concise overview of the informed decision-making techniques includes informing science approach, and tradeoff optimization methods, and focuses on the applications of crisis-resistant knowledge and technology transfer in digitalization. As a concept proof, we further present the case method framework and reinforce the principles and practices introduced by a set of case studies. Therefore, we demonstrate this multi-perspective approach applied to investigate such a complex phenomenon as digital transformation and validate the ITC framework.

Chapter 3 illustrates the pre-digital era transformations that happened in large-scale, complex business environments. The complexity, as a possible crisis trigger, embraces a number of mission-critical aspects such as organizational structure, geographical range, and business diversity, to name a few. According to the ITC-based method, we cross-examine these transformations by case studies. The IT-intensive businesses for our real-world case studies include large-scale multinationals in software-intensive production, IT consulting, and fashion design.

Chapter 4 examines organizational transformations in the digital era, which take place in rapidly changing and highly competitive environments, which complicate business landscapes and may trigger crises. We present these dramatic transformations as real world-based case studies and compare IT-intensive businesses such as publishing, and fast-food production and delivery. For these cutting-edge businesses that include both established companies and recent startups, we discuss franchising and outsourcing options.

Chapter 5 investigates the real-world cases of digital era transformation in large-scale multinationals, such as Microsoft and Huawei. Their organizational complexity is followed by a multidimensional diversity, which includes national, religious, gender, cultural, and a few other aspects. The IT-intensive businesses we survey include telecommunication services, and digital production and delivery. In these challenging environments, the transformation strategies critically depend on managing technology, knowledge, and customer relationships.

Chapter 6 discusses digitalization in the forest industry, an essential contributor to the Russian budget. Being traditional in Russia for many centuries, this business area features stiffness, which complicates its digitalization. However, the new and emerging environmental requirements call for agile transformation in order to survive. To optimize the solution, we examine a set of tradeoffs that includes such potential crisis factors as organizational bulk, complex logistics and legislation, and local and national diversity.

Chapter 7 examines the digitalization challenges originating from miscommunications between the client and the developer of the digital product. These misconnections are crisis triggers, the root cause of which is the so-called "human factors". Although often neglected by the management, these may appear mission-critical for digitalization and sustainable development. Being one of the "three pillars" of the ITC framework, these include such valuable "soft skills" as teamwork, negotiation, and self-criticism. Therefore, properly addressed, these typically become mission-critical drivers for successful digitalization and sustainable development.

Conclusion summarizes the research outcomes of the book and suggests perspectives for the post-digitalization society and business development.

References

1. Church, A. (1941). *The calculi of lambda-conversion* (p. 92). Humphrey Milford Oxford University Press.
2. CMMI. Retrieved from http://wikiitil.ru/books/CMMI-07.02(rus).pdf (Last Accessed February 17, 2022).
3. Gill, T. G. (2011). *Informing with the case method: A guide to case method research, writing, & facilitation* (p. 563). Informing Science Press.
4. Schönfinkel, M. (1924). Über die Bausteine der mathematischen Logik. *Mathematische Annalen, 92*, 305–316. Retrieved from https://doi.org/10.1007/BF01448013 (Last Accessed February 17, 2022).
5. Zykov, S. (2016). Adding agility to enterprise process and data engineering, Corpus ID: 30556519. Retrieved from https://www.semanticscholar.org/paper/Adding-agility-to-Enterprise-Process-and-Data-Zykov/50a96c062efaa2b1941e498f3121661 12dac4d37 (Last Accessed February 17, 2022).
6. Zykov, S. (2018). *Managing software crisis: A smart way to enterprise agility.* Retrieved from https://www.semanticscholar.org/paper/Managing-Software-Crisis%3A-A-Smart-Way-to-Enterprise-Zykov/67c9e60df5dfceecb2ff746eb8585043d926b072 (Last Accessed February 17, 2022).
7. Zykov, S. V., & Singh, A. (2020). *Agile enterprise engineering: Smart application of human factors. Models, methods, practices, case studies.* Springer Nature.
8. Zykov, S. V. (2021). *IT crisisology: Smart crisis management in software engineering models, methods, patterns, practices, case studies.* Springer Nature.

Chapter 2
The Case Method: Promoting Informed Digitalization

2.1 Digitalization in Russia: Contemporary Survey

The term "digital economy" has become more and more frequently used in our lives: the media mention the digital economy "illness" of the Russian president, researchers devote articles to the discussion of this term and its components, industries invite experts to assess the current situation of the market and its future development prospects. Therefore, an interest in changes has grown. What are the roots of the changes? Do the changes really deserve attention?

The electronic relations have appeared by spreading widely the new technologies such as the Internet all over the world, particularly in Russia. The huge variety of opportunities, such as new tools and resources, is provided by a pervasion of different life spheres, such as economy, by electronic relations. These changes have led us to the Fourth Industrial Revolution (Industry 4.0) that covers the whole world and becomes a new level in the development of the world economy. There is no doubt that the Industry 4.0 will modify people's lives like electricity did. It is necessary to follow and support the evolution of the electronic relations to gain the leading position in the world.

The digitalization has become a trend in the world because it broadens the horizons of common things and allows developing new and smarter approaches. Consequently, industries can increase income and customer's satisfaction, and customers can make the services available easier/better or obtain completely new ones. The government optimizes the processes of providing the municipal services via Internet such as (1) electronic registration; (2) electronic queue; (3) the electronic distribution of social events; (4) electronic participation in these events. Therefore, people can (1) save their time; (2) provide/obtain better/easier services; (3) promptly get informed on the social events and participate in the preferred ones. To make a long story short, the digitalization develops the nations and economies for the benefit of the people.

In 2016, the President of Russia, Vladimir Putin, proposed the plan of digital economy development [16]. In 2017, the Russian Prime Minister Dmitry Medvedev approved the national program of the digital economy. Both events demonstrate

© The Author(s), under exclusive license to Springer Nature Singapore Pte Ltd. 2022   7
S. V. Zykov, *IT Crisisology Casebook*, Smart Innovation, Systems
and Technologies 300, https://doi.org/10.1007/978-981-19-2231-2_2

the Government's involvement in economy digitalization and national development. Such modernizations may indicate the upward trend in competitiveness and national security compared to other countries. Therefore, digital economy development positively affects the lives of both individual citizens and the country.

Let us analyse the development process of the Russian digital economy including its constraints and limits. Our scope will include the following:

1. Analyse the history of digital economy.
2. Define the concept of digital economy.
3. Analyse the current state of digital economy.
4. Detect the braking and pushing factors in the development of the digital economy in Russia.
5. Appraise the Government's role in the development of the Russian digital economy.
6. Recommend activities and strategies for the Government and industries.

The objects of this research are hi-tech companies and countries/regions actively involved in the development of digital economy.

Digitalization of economy is critically important in country development. What is the state of economy progress in Russia? What is the position of the Russian economy in the world?

The study of the trends and experiences can assist in answering the above questions. Analysis of braking and pushing factors can reveal the weaknesses and strengths in digitalization. Such analysis allows faster detecting the critical issues and mitigating/preventing them. The result is a recommendation list for the Government and industry.

To avoid terminological ambiguities, let us define the following terms:

Digital economy is a system of economic, social, and cultural relations based on digital information and communication technologies [9].

Digitalization is using digital technologies in business to radically improve its performance [3].

Industry 4.0 is a massive application of cyber-physical systems to industrial production including 3D printing, big data analytics, Internet of things, virtual and augmented reality, and the related technologies [9].

Information and communication technologies (ICT) is a set of methods, production processes, and software and hardware integrated with the purpose of collecting, processing, storing, distributing, displaying, and using information for the benefit of its users [15].

The digital divide (or *information inequality*) is a disparity in accessing ICT that results in growing economic, social, and cultural inequalities [26].

The first step of this research includes discussing the concept of digital economy and identifying its global trends and effects. The key idea is to determine the current state of economy based on industry cases that reflect the current trends and effects of digital economy.

The outcomes are

(i)   List of pushing/breaking factors of the development of the digital economy.
(ii) List of digitalization strategy recommendations for industry and Government.

Theoretical background

Multiple source data analysis suggested that the objective is twofold: (1) under-standing the digital economy development in the world, and its general trends; (2) exploring the aspects of the Russian digital economy. To accomplish these tasks the reports of leaders in consulting were investigated, such as Deloitte Touche Tohmatsu, Ernst & Young, KPMG, PricewaterhouseCoopers, McKinsey & Company, Accen-ture, Boston Consulting Group (BCG), and AT Kearney. All these regularly invest in digital economy research and strive to support digitalization development. In addi-tion to these sources, the proceedings of the World Economic Forum and a few white papers by IT-intensive companies were also reviewed.

Studying digital economy requires exploring the "Industry 4.0" as these concepts are indivisible. In 2016, PricewaterhouseCoopers published the Global Industry 4.0 Implementation Review. According to the company, this review was the largest research of its kind. It included such key points as the meaning of the term, Industry 4.0 features, important digital technologies, and practically focused outcomes.

In 2017, the McKinsey consulting company presented their "Digital Russia: A New Reality". The report describes in detail how the digitalization process in Russia can increase GDP growth, what sources for this exist, and the potential outcomes by 2025. The authors emphasize that such a development path promises to boost the Russian economy and individual lifestyles and that the ambitious goals set are potentially achievable. They highlighted the need for and importance of cooper-ation between the Government and businesses. In addition to information on digi-talization in Russia, the report provides information on the impact of the digital revolution around the world.

Accenture's 2016 report is devoted to finding the critical factor for winning the digital race. According to experts, the human factors will trigger the digitalization success. The paper also stresses that many companies are exhausted by the race, trying to keep up with the competitors. As a result, the business environment faces a "digital shock". The remedy is focusing on human factors, specifically on the customers, so that the innovations to be implemented eventually meet their requirements. The report also emphasizes that being customer-oriented is necessary for survival.

Another company, BCG, argues that by 2035 the digital economy could reach $16 trillion. Their report also describes the expected positive effects of technology progress in several countries. When analysing the situation in Russia, experts empha-size that the lack of a harmonious interaction of all participants in the digital economy greatly hinders its development. Therefore, they propose using the experience of the UAE and Saudi Arabia and developing the public services, online education, and medicine. Also, analysts appreciated the current growth of Russian exports, which contradicts with McKinsey's pessimistic predictions. Nevertheless, their experts agree that digitalization is a possible source of many promising prospects.

The history of the digital economy

Where does the digital economy came from?—This is a critical question for many people as it is the information that often changes the labour market, depriving people of certain professions. Prior to Industry 4.0, the world experienced three industrial revolutions: the first was marked by the transition from manual labour to machine labour, the second—by the transition from manufacture to the factory, and the third—by the automation of processes. Each of these revolutions wiped out certain professions and many people became unemployed. Today, it is difficult to imagine a world without the new achievements resulting from these revolutionary changes, and hard to imagine how the previous generations resisted the innovations that later transformed and eased their lifestyles. Indeed, many of them did not lose their jobs but were engaged in the new activities, more creative and less labour intensive.

Currently, the world is absorbed by the next revolution, Industry 4.0. Perhaps, the year 2017 marked the point of no return, as 50% of the world population became Internet connected [9]. According to the McKinsey Global Institute's forecast, in 20 years the effect of the fourth industrial revolution will be comparable in scale to the industrial revolution of the 18–19 centuries, which halved the number of the people employed in the primary sector of the world economy. Back then, this process required over 160 years, now it can happen 8 times faster. As such, it is likely that the revolution will automate nearly half of the processes.

The fourth industrial revolution, also referred to as "Industry 4.0", is the digitalization of all physical processes and their integration into the ecosystem involving the entire value chain. Figure 2.1 shows the technologies which drive digitalization [3].

The term "digital economy" deserves special attention. The appearance of the Internet in 1982 triggered the advent and further development of the virtual world.

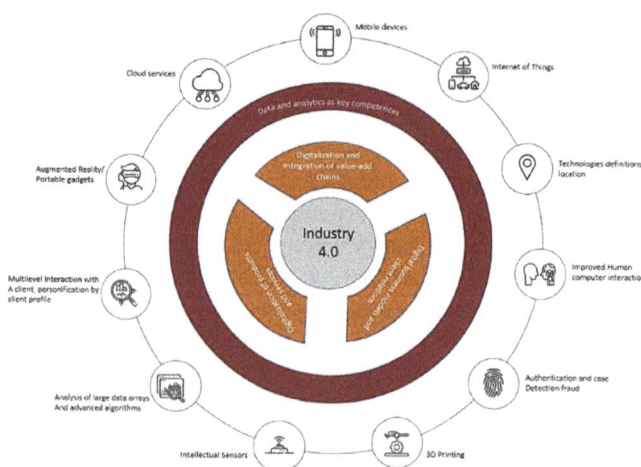

Fig. 2.1 Key digital technologies

New games, social networks, tools, forums, and the like appeared online, attracting users with their usability, and engaging them more and more into this virtual world. Whereas it was possible to determine the boundaries between the real and the virtual world, these have now practically disappeared. A new space has appeared with its own laws and rules, of which the economy is a part—the digital or virtual economy.

This term was coined by a computer scientist from the United States, Nicholas Negroponte, over 20 years ago. He presented the advantages of this new economy: lower importance of physical weight and boundaries, lower production costs, and gradual replacement of some material goods by their digital copies [26]. In the Information Society Development Strategy, this term was presented as: "… economic activity, in which the key factor of production is data in digital form, processing large volumes and using analysis results of which, compared to traditional forms of management, can significantly improve the efficiency of various types of production, technology, equipment, storage, sale, delivery of goods and services" [15].

The pace of changes caused by digital economy is impressive. While the transition from older generation computers to personal ones took several decades, digital economy brings revolutionary changes within years or even months.

The global economy is shaped by accelerating waves of innovation. The first wave was marked by process automation, the second wave was triggered by the massive proliferation of technologies such as the Internet and mobile communication. Today, businesses have realized the potential economic and social benefits of digital technologies (Fig. 2.2 shows some of these) and are therefore looking for the ways to achieve these benefits. Therefore, the current wave can be described as "changing the business operation model". This process is especially intensive in the financial and banking sector: a dramatic example is the creation of the Tinkoff Bank, which is radically different from many traditional banks. Tinkoff's example proves that the new operating models not only allow them to break into the established market but also increase their efficiency to the degree that the other market players adopt their experience to stay competitive.

It may seem that innovations are not capable of changing traditional industries, such as forest industry, agriculture, and manufacturing but this is far from being the case. One such example is the Sibur digital plant where production line sensors detect

Fig. 2.2 Benefits of digitalization

the bottlenecks and optimize the production [22]. To sum up, digital technologies apply to a wide variety of industries, generate innovative solutions and improve management and production efficiency.

The impact of the digital economy

Digitalization has a positive effect on the development of countries. As the purchasing power of the population grows, goods and services become more accessible, and the competitiveness increases, which again stimulates the market expansion up to the entire sectors of the countries and internationally. In short, digital transformation, provides new methods and tools to achieve economic benefits and ensure sustainable development of national economies.

In addition to advancing the nations, digitalization provides new prospects for each individual: people can access new knowledge, grow professionally, and expand their horizons easier, faster, and better. The absence of geographical barriers boosts these processes. Daily life also improves the public services to become better accessible and actively involve the citizens into the city improvement to make their living conditions more comfortable and increase efficiency of the governmental organizations. Countries striving for change and innovation are most attractive to qualified personnel. In its turn, the growth of professionals and experts stimulates further development of the economy.

For the companies, digital technology is the key to success and stable development. The companies ignoring innovation become outsiders. A vivid example is the Airbnb company established in 2008: it currently offers three times more options to its customers than the three largest hotel chains known for many decades.

Development of digital economy drives customer purchasing power. The reason for this is the increased price competition which boosts digital platforms and marketplaces. For example, taxi rides have become much cheaper with mobile applications, such as Yandex.Taxi and Uber. The Yandex.Market marketplace does not only locate cheapest products, but also displays the product features, tracks similar options, compares them, presents the customer reviews, etc. These innovative options, in turn, force the manufacturers to adjust the prices and monitor quality. Also, some services became free, e.g., GPS navigation and cloud storage.

One of the controversial aspects of digital economy is the issue of new opportunities. On the one hand, there is a threat of job cuts, because up to 50% of the processes are planned to be automated. As a result, the number of jobs for people with moderate qualifications will decrease, and the wage gap will increase. On the other hand, technologies provide new employment opportunities: people can independently master various professions and work remotely; geographical and social boundaries are erased. Also, radically new digitalization-related professions emerge. For example, some 20 years ago, an innovative profession of evangelist, i.e., an expert in the company's ideology and product development, appeared [6].

Digitalization is also able to improve the life of every individual citizen, because thanks to it, the quality, availability, and convenience of services, such as educational, medical, social, and others, substantially grows. As a result, the population solves social and financial issues more effectively. Technologies can provide a higher level of comfort and safety: navigation applications simplify moving around and improve this process. Other tools optimize resource consumption, such as electricity. Digitalization assists in automating such processes as sorting garbage, cleaning, and fire suppression. Moreover, technologies allow efficient organizing of the infrastructure and, consequently, improve the lifestyles while keeping within the budget. Such attractive conditions provoke an influx of skilled professionals as these prefer decent living standards. In this case, investment climate also improves thus forming better conditions for further business development.

Another important factor is security. Digital technology simultaneously affects the three levels: the public, the national, and the global one. Currently, technologies improve safety as they assist in reducing the level of crime, preventing fires and disasters, fighting corruption and terrorism, etc. Therewith, digitalization contributes to the fight against fraud. McKinsey experts also believe that an adequate degree of digitalization likely ensures appropriate national security at the state level.

Key digital trends

Digital technologies influence companies, nations, and customers. Each of the above has its own specific trends.

For the companies, it is critical to stay competitive. According to McKinsey experts, digital technologies determine business sustainability. Digitalization allows for optimizing costs, improving business process efficiency, increasing profitability, and better understanding customers; all of these contribute to competitiveness. One of the fastest growing assets of the digital economy is digital platforms. These provide customer interaction, transaction marketplace, and innovative business modelling. Many companies invest in digital platforms, among them are Facebook, Amazon, VKontakte, and Airbnb. An impressive example of using a digital platform is the Russian Yandex.Market, where a user can not only find promising options, but also compare them and pay for them instantly. Moreover, the customers leave "digital traces", i.e., the marketing data to be analysed further. Traditional industries also use digital components: in the industry, equipment has sensors that collect information for further optimization that ensures efficient operation.

However, in some industries introducing technology it is not enough as leadership is also required. This applies specifically to digital-related industries; the "winner takes all" principle governs there. The introduction of digital technology intensifies competition. As an example, let us consider Tinkoff Bank, which in the recent ten years became one of the world's largest independent banks. The reason for this is the fact that easily scalable IT platforms combined with low costs provide dramatic growth rates, which allow quick capturing significant market shares and leaving many competitors far behind.

Another trend is combining conventional ventures and digital platforms. This trend is particularly noticeable in tourism: users prefer digital platforms, where they can compare options, optimize choices, and save time.

The level of specialization of companies is also growing. Companies prefer to increase efficiency by subdividing complex business functions into smaller tasks. As such, new services appear, for example, combined food delivery from various restaurants. In addition to new services, new markets emerge. Amazon, initially a paper bookseller, created a new platform to offer e-books. Further, they added products for smart home and other electronic customer goods. Thanks to innovative technologies, the number and diversity of new markets increases including unmanned vehicles and cloud/e-commerce platforms offering products and services to the customers.

Concerning current business trends, competition has become stiffer and sustainable development requires regular monitoring potential threats of emerging competitors/services and continuous striving for the digital leadership.

Recent technological innovations increase efficiency of the budget funds that improve infrastructure, healthcare, and education.

Another important trend for the customer is the change in the labour market. The changes are not only positive: some companies prefer to save by cutting jobs, and new job offer numbers decrease. Average qualification employees are especially affected as process automation reduces the need for them. Nevertheless, the positive aspects exist: new professions appear, the approach to education changes, and the elimination of geographical boundaries provides wider opportunities for finding jobs and adding qualifications.

For customers and companies, not all trends seem positive. However, the innovative technologies tend to increase the quality of life.

States also have specific trends: one is providing public services exclusively in digital form. Thus, the option of "bypassing the system" is excluded, and the public services centres change for citizen assistance centres. Moreover, convenience of interaction through digital platforms creates new principles of work, such as individual approach, single-space services, and quality support. One example of this approach is the Danish NemID service, the single channel combining governmental, banking, and some other services.

Another trend is partial automation of governmental processes and their integration as a single space. This also boosts the activities of the citizens when interacting with governmental agencies, thereby increasing interaction efficiency and level of citizen satisfaction. Special attention should be paid to the mechanism of publishing data for universal access; thus allowing access to rich data sets, such as budgets of state institutions. Here, the trend for the government agencies is their increasing use of analytical methods and data analysis. This improves decision-making quality and, consequently, more efficient decisions.

Most key governmental digitalization trends turn out to be positive as they generally contribute to work efficiency and client satisfaction.

Success factors

Introducing digital technology is not only a complex process, but also an expensive one. Sometimes it may even seem that the results do not pay off the investment. Businesses tend to use cheaper labour to stay profitable; management often takes employee social obligations for granted. Therefore, it is very challenging for businesses and nations to make decisions regarding technology deployment.

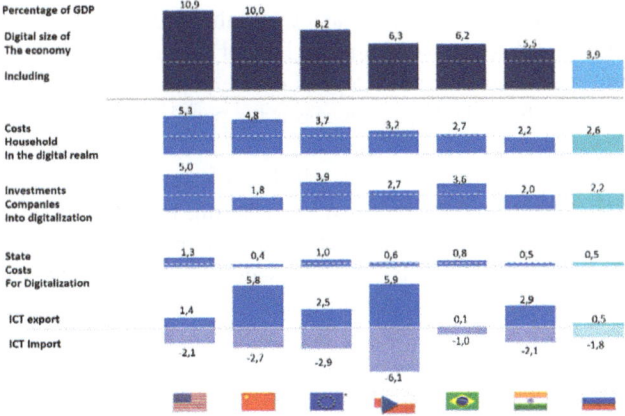

Fig. 2.3 Digital economies and GDP

However, one of the success factors is digitalization costs embracing individual households, companies, and entire nations. Figure 2.3 demonstrates that the leading countries, such as the USA, China, and some EU states, occupy front-rank positions in terms of expenditure, which correlates with the share of their digital economies in GDP.

Experts note that the benefits of innovative technologies are already obvious, which motivates companies to invest in these areas. Adopting this fact and developing ambitious goals based on innovative advancements is often a key to sustainable development for private businesses and entire industries and economies. As a result, companies manage to get extra profits, save on operating costs, and create new products, services, and progressive business models.

Increased productivity is one of the most important benefits companies can get. However, rather than choosing between increasing gross revenues or profits, it is possible to improve these two indicators together. According to the forecast of PricewaterhouseCoopers, companies can get up to 2.9% more revenues and reduce costs by 3.6%. For those who have made bets on breakthrough innovations, the results are planned to be even more impressive. In any of the above scenarios, the cumulative costs decrease worldwide, and the total annual revenues increase.

Changes of the businesses are largely determined by their customer needs. Here, along with the digital technologies, a new opening is efficient customer interaction and related business process transformation. As a result, products and services increasingly meet the customer expectations, which stimulates further development of companies. Big data analytics can drive such a progress; however, creating digital platforms brings the innovation leaders to even more promising positions. In short, customer focus is the key factor to succeed in development.

The influence of the human factor does not end there as companies understand that digitalization is most effective when involving employees and establishing an adequate corporate culture. When employees neglect the transformations, this hampers the entire process. Another factor of value is the percentage of internationally certified employees. According to Digital IQ, technology investments are important; however, their outcomes depend on the people who actually implement them [17, 20]. Therefore, the key to success is the people who perform dynamically in digital ecosystems.

Yet another success factor is confidence in digital solutions. For example, business analysts are no longer directly involved in gathering the data and assessing their value. Rather, they assist in decision-making to ensure adequate judgment quality. Utilizing complex enterprise software, harnessing risk management analytics, and ensuring data reliability may dramatically improve the information quality. Realizing this, many innovative companies create dedicated departments responsible for data analytics.

Ironically, investment remains a key success factor. This is equally applicable to technology, education, and organizational transformation. Over 50% of the companies surveyed expect the investment to add income in the two upcoming years.

Mid-conclusion

Obviously, digitalization has already significantly changed companies, industries, markets, and economies. The impact that it has on all participants is generally positive. The market grows and is likely to grow in future. The technological prospects will probably improve the world dramatically. Although certain digital economy trends such as increased competition may seem negative, these factors also have positive effects. Most digital trends are positive for individual households, private companies, industry sectors, and entire nations.

Digital economy opens innovative development prospects for all its actors.

Analysing Russian digital economy

This section presents the evolution of the Russian digital economy including past, present, and future activities.

According to McKinsey experts, Russia has already entered the digital era. By the number of Internet users, Russia ranks first in Europe and sixth in the world. 60% of the world population use smartphones; a few years ago, their number was twice less. By this indicator, Russia is ahead of Brazil, India, and the countries of Eastern Europe. The number of Russians using state and municipal service portals is increasing rapidly: it recently exceeded 40 million [9].

The experts expect the economy digitalization to be one of the key factors for GDP growth. According to McKinsey, by 2025 the digital economy size is going to reach some 19–34% of GDP. This effect is planned to be achieved by automation, business models, and breakthrough technologies including 3D printing, big data analytics, and robots. For example, until 2025, the Internet of things only is expected to generate a total profit of $4–11 trillion USD worldwide.

Compared to GDP, the digital economy is expected to grow several times faster. Russia's digitization has grown significantly in recent years. The examples of this growth are such successful businesses as Yandex, one of the top 30 global companies by revenue [14], and VKontakte social network, a leader by the number of users in

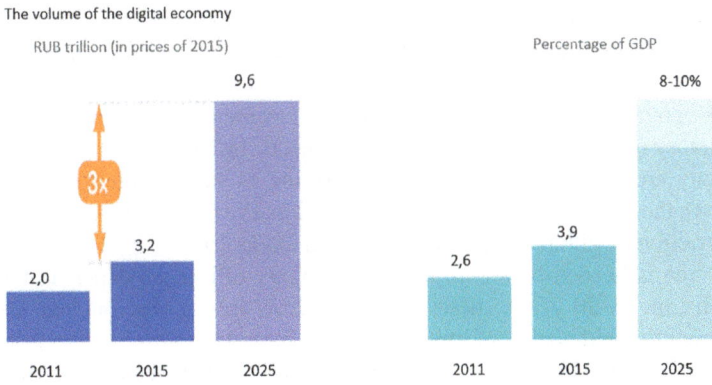

Fig. 2.4 Digital economy size, trln RUB

the former USSR, which transforms to a full-fledged digital platform offering a rich variety of goods and services. The labour market also improves; major infrastructure projects are being implemented. Nevertheless, the gap between the volume of digital economies in Russia and the leading countries remains significant. Reaching their level seems unrealistic, while tripling the digital economy size looks an achievable and meaningful goal (see Fig. 2.4). After implementing this, Russian digital economy size becomes comparable to that of the leaders; however, further growth will not be as intensive.

So far, Russia is behind the leading countries in a number of indicators: percentage of people purchasing online; internet and smartphone penetration; percentage of website enabled organizations, etc. However, the key indicator is the ratio between the economy size and the aggregate GDP, which is 2–3 times lower than that of China or the USA. Digital expenditure indices are also significantly lower, both for the individual households and public/private investments. The export level of digital technologies is four times lower than the import level.

Legal regulation of Russian digital economy

For Russia, as for any other country, digital economy is an important ingredient of the development strategy. Russian President Putin highlighted digital economy as the main strategic direction of the country's development [16], realizing its advantages such as GDP growth, increasing world competitiveness, solving national security issues, etc.

Russian executives developed a set of legal acts to support digital economy development. One of these is the National Technology Initiative (NTI), a long-term comprehensive program to set up leadership conditions for the national companies in the innovative high-tech markets. The program was launched in 2014 with its main task to launch smart high-tech projects. By 2017, 32 of such projects were approved, and 57 priority ones were identified. Also, twelve markets and directions of NTI were formed, and eight road maps were approved. The key performance indicators for 2018 include establishing a base of 10,000 companies and an expert pool, survey-based interaction with 1000 companies, and priority support for 100 of these.

The program demonstrates the Government's awareness of the need to improve the regulatory framework for the innovative projects. In 2017, seven workgroups for key activity areas were established to identify the existing administrative barriers and remove these by means of improved legislation. For each of the workgroups, the legal restrictions highlighted are summarized in Table 2.1.

Each area is headed by a competent technology officer, an expert in the field. Expectedly in 15–20 years, these areas will grow into global markets ensuring national security and quality of life.

In 2016, President Putin instructed the Russian Prime Minister, Dmitry Medvedev, to create the national digitalization program. In 2017, Medvedev said that the program was created and approved by the Order of the Government of the Russian Federation. The "Digital Economy in the Russian Federation" is the key program document. According to Medvedev, this document was intended to "organize the system development and introduction of digital technologies in all areas of life". The program authors suggest the project strategic outcome as a full-fledged environment to support efficient development of the national digital economy. However, this process requires acceleration to catch up with the leaders.

The objective of the program is threefold:

1.   Creating the Russian digital economy ecosystem.
2.   Creating the institutional and infrastructural environment.
3.   Increasing the competitiveness of national economy and its industries.

For the program to succeed, multi-level transformation is required. The program highlights the following three:

(i)   Markets and economy sectors.
(ii)   Platforms and technologies.
(iii) Environment.

The main development spheres include the following five components (see Table 2.2):

The other application areas are:

1.   Public administration.
2.   Smart city.
3.   Health care.

The program also lists the end-to-end technologies that will allow the industry to move to a new technological structure:

- Big Data.
- Neurotechnology and artificial intelligence.
- Distributed registry systems (blockchain).
- Quantum technology.

Table 2.1 Workgroups on the Russian digital economy

Workgroup name	Objective	Constraints
AutoNet	Distributed network for self-driving cars	1. Lack of conditions to assess the effect of innovations on the international market 2. Lack of legal regulation
AeroNet	Distributed network for self-driving aircrafts	1. Lack of operator certificates 2. Lack of space for flights below 150 m
MarinaNet	Smart management systems of unmanned maritime transport and ocean development	1. Lack of open and available geoinformation data 2. Lack of legal framework for autonomous tools
NeuroNet	Distributed artificial components of consciousness and mind	1. Lack of legal framework 2. Lack of technical regulation for product quality assurance
TechNet	Smart and innovative technologies	1. Lack of legal framework 2. Lack of recognition and adoption of computer-based experiments
HealthNet	Personalized medicine	1. Legislation improvement based on the regulatory principles of medicine circulation 2. Need to develop the medical care quality criteria for outpatients
EnergyNet	Distributed energetics	1. Lack of legal framework 2. Need to eliminate regulatory barriers and integrate the consumers into a single power grid

Table 2.2 Digitalization spheres of the Russian economy

Key sphere	Objectives
1. Regulation	Establishing favourable legislation to support digital technology development and implementation based on a sustainable regulatory environment
2. Personnel and education	To develop digital economy, technologies and further education should be available/accessible. This requires qualified IT professionals, improved education system, and advanced labour market
3. Research competencies and technology background	Applied research in digital economics is required to support national security and competitiveness. This requires a reinforcement framework
4. Information infrastructure	Establishing individual communication networks, data centres, digital platforms, and other infrastructure elements
5. Information Security	Ensuring safety for the citizens, data, and infrastructure

- Industrial Internet.
- Robotics and sensor components.
- Wireless technology.
- Virtual and Augmented Reality.

The document includes key objectives and directions for the digital economy development, and a roadmap. The development program is until 2024 and assumes that all its KPIs to be achieved by this deadline. This document considers the factors hindering digital economy development to eliminate them and create favourable conditions for digitalization. For instance, as consumers are not yet accustomed to technology, one of the tasks is building their "digital awareness". The regulatory framework requires major improvements and plans that in 2018 the necessary legislation appears to support the digital economy [19]. The program also suggests compensating the shortage of qualified IT staff by annually increasing their number by 120,000 starting 2024. This will contribute to resolving the digital illiteracy issue. In particular, the share of the "digitally skilled" population is expected to grow up to 40% [5].

Introducing digital economy, it is important to remember that small and local changes are unlikely to result in adequate returns. The implementation should be comprehensive and address both smart innovations and existing barriers. The developed program generally meets these requirements and aligns well with the related national strategies.

Digital inequality

One of the factors hampering the development of the digital economy in Russia is the problem of digital inequality. Selischeva argues that "The Russian economic environment is characterized by a great heterogeneity and uneven development" [20]. Indeed, certain Russian regions such as Moscow and the Tyumen rapidly introduce ICT taking advantage of innovations, while some other regions with insufficient Internet coverage drastically lag behind [24].

Information environment heterogeneity hampers efficiency of the digital economy development; however, the root cause of this problem is insufficiently studied. Nevertheless, the problem of the "digital divide" between the regions is particularly acute in Russia, and therefore requires a solution. We identified five factors that influence this inequality.

1. High cash threshold.
    First, the digitalization costs are substantial for the state. Second, for some Russians ICT using is not feasible due to high cost of living. This problem results from economic inequality. Russia is a leader in social stratification level [27]. Low-income citizens have a limited access to high quality education; they are often deprived of mastering ICT. Also, as previously mentioned, some Russian regions do have an insufficient level of Internet penetration; this increases digital inequality.

2. Undeveloped, poor quality, and expensive ICT infrastructure.
    Russian telephone lines do not always meet Western standards, telephone density is low, and connection speed in certain regions is 10 times less than that in Moscow.

3. High prices of Internet providers.
    Internet providers offer services at high prices. For instance, in 2010, an average Khabarovsk citizen, paid 1,700 rubles (approx. USD 24) for Internet access, which accounted for 18% of his income [4].

4. The degree of purposeful Internet use.
    This refers to the goals of using Internet, e.g. for entertainment and study. While Russia ranks second in the number of Internet users, the country is only number 53 in terms of the culture and intensity of its use. These ratings are comparable to Brazil, Indonesia, and India, while the digital leaders such as UK, Sweden, and Denmark are far ahead. Considering the situation in Russia, one should also note a gap in the purposeful use between Moscow and the regions: the capital is 2.4 times ahead of the average [2].

5. Size of the country
    As the area of Russia is 17.2 million km^2, additional costs are required to cover the territory with the communication lines. The cost of information products and services varies greatly depending on the region: the rural areas require higher ICT infrastructure costs.

6. Globalization
    The ICT enabled people to enjoy advantages of job hunting online including abroad offers, and better labour mobility. Internet allows small and medium businesses to improve their infrastructure by using advanced services such as payment systems. The boundless Internet allows business cooperation with

foreign companies and customers. In short, advanced ICT users get more benefits than the people without ICT access and as a result, the digital inequality grows.

7. Undeveloped regulatory framework.

New relationships arising from the use of ICT require new rules and restrictions; however, Internet regulations are often insufficient. The state should more actively contribute to the national ICT development.

Obviously, the digital divide has a number of negative consequences, one of which is an unfair competitive environment that disrupts stable development of the economy as a whole. Technologies change the society by creating an information culture that promotes the ICT interaction and stimulates economy development. The digital divide demolishes this advantage and reduces infrastructure efficiency.

Success factors in Russia

First, let us note that the ICT infrastructure level in Russia is good enough compared to other countries. Internet tariffs in Russia, although higher in certain regions, are still about half the same in the EU [12]. Moreover, the Internet costs including mobile tend to reduce. The prevalence of mobile broadband access is about 60% [1]. The Government also promotes Internet availability and expansion, the evidence of that being wi-fi availability growth in Moscow subway and ground transportation. Internet bandwidth is around 12 Mbit/s i.e., above average, which puts Moscow ahead of the capitals of Brazil, China, Italy, and some others [8]. The above facts indicate an increase trend in digital service availability.

Importantly, Russia was among the first to launch R&D on the new generation of 5G Internet with a bandwidth of 10–20Gbit/s. This paves way for smart and innovative digital technologies, such as more detailed analytics based on IoT devices connected to homes, cars, and production lines.

However, the factors stimulating digital economy development are region dependent since the level of infrastructure availability varies greatly from one area to another. The leaders are Moscow and St. Petersburg, whose levels are similar to those of the leading countries, while in such regions as Khabarovsk the average bandwidth is four times lower and 40% more expensive [13]. A positive point is the Government's awareness of this problem and the desire to solve it by the program for digital equality.

Another important indicator of digital economy development is the distribution of the expenses on the digital technologies between the state, businesses, and households. Unfortunately, these indicators in Russia are several times less than those of the leaders, USA and China (Fig. 2.1.3): the Russian household expenses make up to 2.6% of GDP, the businesses contribution is 2.2%, and the state one is 0.5%.

Another factor positively affecting Russian digital economy is high e-commerce growth rate. In the past few years, online retailers started offering goods and services at lower prices and shorter delivery time than traditional businesses. Also, new delivery options appeared such as post-machines and specialized delivery terminals.

Delivery options became more comfortable due to prevailing tendency to return goods upon receipt, and a variety of payment methods such as online, cash, and card. The purchasing power that decreased in crisis provoked Russians to explore new markets, which triggered e-commerce development.

Recently, Russian online customers paid more attention to foreign traders as Chinese companies and trading platforms were probably the most attractive. As a result, they captured a substantial market share: according to Association of Online Trading Companies, in 2016 international shipments increased by 102%, while domestic grew by only 6%.

Active digitalization of individual consumer markets, among other things, is a stimulating factor for development. Internet banking turned out to be the most intensively developing area; according to Finalta, their level has reached 30% in five years [23]. Digital platforms stimulated development of service-oriented industries; vivid examples are taxi services—Gett and Yandex.Taxi, personal services—YouDo, and online cinemas and music. However, there is a threat of foreign penetration and capturing a larger market share. Such international companies as Google, Uber, and AliExpress are extremely popular among users and can really crowd out domestic companies. For the latter ones this is a real threat, while for customers it is an option to purchase cheaper, with better comfort and satisfaction.

In the medium term, a high growth rate of e-commerce is expected; it is this factor that should bring Russia closer to world leaders. The ways to promote these include strategies for conquering crises, gradual elimination of digital inequality, and increasing adoption of smart and innovative technology.

Private Russian companies lag the leading countries in digitalization. This is due to their lower hi-tech involvement and poor investments in productivity, innovative products, and services. In the domestic market, the Russian companies are exposed to stronger foreign competitors. Clearly, in the international market their competitiveness is even less. The situation is aggravated by low customer involvement as the local market should preferably stimulate domestic digital solutions.

As a result of the lack of an adequate level of investment, many sectors of the economy remain far behind world leaders, including mining and processing industries, which are key sectors for Russia. Digital technologies provide a set of methods and tools with which it is possible to realize the potential that has not yet been revealed, but it is necessary to act quickly. Nevertheless, there are industries whose level is as close to the world level as possible, among them: education, finance, ICT.

Another factor hampering digitalization is a business reluctance to certain technologies, particularly, process automation by ERP and CRM information systems. According to official statistics, only 10% of companies use such systems. Although certain systems are not included in the official statistics, technology adoption rate remains extremely low. Most likely, the reasons for this are reluctance of SME to develop such systems and slow implementation due to low expertise and resources. Information systems should be actively involved in production; the businesses should strive to implement the breakthrough innovations. Otherwise, they are at risk of competing with radically innovative players, which is doubtful to conquer. Concentrating resources and using standards will help to overcome such business model gaps.

Strategic partnerships also open new prospects. For instance, Tinkoff Bank demonstrated how business cooperation increases the level of competitiveness: their cooperation with the BCS broker boosted new clients for both parties, and for the bank it also assisted in launching their own innovative brokerage service [21, 25]. Partnering with businesses, governmental and R&D organizations significantly increase maturity level. As such, experimenting with innovative technologies, choosing most efficient, and discarding the deprecated ones assists in sustainable development. Obviously, investment level affects digitalization prospects. Companies need to revise the investment policy, and the Government should create favourable conditions for private investments.

The next factor which influences digital economy development in Russia is low share of exports of digital goods and services (5 times less than the leaders) and high level of their imports. Rapid development of the world software market provokes users to pay attention to foreign products. The bulk of exports contributes to custom development rather than licensed software, which brings little profit to Russian companies and causes further foreign client growth. Also, the rates and volumes of software export growth are much lower than those of other exporters, even non-leaders such as Poland and Israel. The trend of moving head offices of large IT companies also impacts negatively, although they leave large development centres in Russia and support IT education. Russia's dependence on imports in certain market segments is critical: about 75% of software and 90% of equipment are imported [19].

It is necessary to increase the volume of exports and reduce the volume of imports. Developing the potential of ICT will make it possible to achieve these goals. To promote macroeconomic conditions, the state chose the policy of import substitution forcing businesses to use domestic software and providing tax incentives for IT companies. Another important step is moving from custom development to licensed products and services with high added value. With the governmental support and IT education development, Russia has every chance to advance its export potential. To support domestic producers, the Government imposed the foreign companies a set of restrictions. These included VAT for all digital content providers, obligatory registration of top online stores as tax agents, and strict customs clearance procedures for express parcels. With such constraints, the Russian customers turn to domestic digital companies. Cross-border trade in many regions is particularly important. The benefits provided promote local business development.

In addition to existing companies, stimulating new startups is required, although Russian extremely unfavourable environment is another negative factor. So far, these were few; however, the major players are likely to become innovation leaders. These include Sberbank, Tinkoff Bank, and Kaspersky Lab implementing large-scale innovation programs. To increase the number of companies engaged in innovation, not only major players, but also a few startups should get support. Compared to the USA, their current growth rate is relatively low.

Digital economy development requires effective venture capital market, currently hampered by financing problems and industry intricacy. Market players' knowledge is often insufficient to invest effectively, entrepreneurs have little experience in high-tech markets; therefore, mentors and investors are not ready to finance business development. This situation is exacerbated by the inability of households to meet the substantial demand for innovation and technology due to low purchasing power. Venture investment in Russia recently shows a negative trend as the businesses prefer ready-made and proven products to innovative ones. In turn, lack of investment resources is the main factor limiting innovative business development. Although state financing level corresponds to that of developed countries, the private one looks depressing: the investment share of GDP is 3 times lower than that of the USA, while annual development rates are 5% only. Unfortunately, there is little hope for private investment to grow substantially as its current amount is just enough to support the market.

Regarding innovation investments, in 2016, the ICT segment accounted for about 71% of their total [10]. Large Russian IT companies invested in startups, one recent success story was Mail.ru Group financing Prisma photo processing service. Accordingly, this market gets a chance to develop intensively.

Intellectual potential is another positive factor for the digital economy: many Russian developers are in demand both in domestic and foreign markets as they have international project experience. The country also has a solid research and education base, which develops through cooperation between the state and business.

Summing up, the potential of Russian digital economy is not fully utilized as the domestic market has a large capacity. Employing the positive factors and addressing the negative ones, given the Government's active support, the goal of tripling Russian digital economy size seems achievable.

RESULTS AND DISCUSSION

Digital economy is able to self-develop independently; however, a persistent nationwide development strategy requires activities to stimulate it.

We have considered the factors affecting the Russian economy digitalization and either promoting or hampering its development, and produced the following recommendations:

- Monitor technology-intensive trends and the threats of new players in the market.
- Strive for innovations and prioritize technologies.
- Analyse partnership options with businesses, Government, and innovation centres.
- Invest in innovations.
- Automate processes by implementing information systems such as ERP, CRM, etc.
- Create innovative types of operating activities.
- Develop technological company infrastructure.
- Develop corporate culture open to innovations.

The study revealed importance of the state role in forming digital economy in terms of creating operational environment and ensuring sustainability and safety. Government agencies should address the above-mentioned factors to establish a favourable business climate that meets the following recommendations:

- Create an infrastructure that provides easy access to competitive project selection and ensures fair competition for the participants irrespective of their location, i.e., simplifies the process of innovative project selection.
- Balance budget burden by multi-level funding of selected startups.
- Create favourable conditions for qualified personnel and business in the country.
- Attract foreign investments by simplifying visa application, ensuring intellectual property protection, and creating innovation centres, business incubators, and accelerators.
- Introduce extra tax incentives for enterprise innovative activities.
- Finalize the regulatory framework for crowdfunding platforms to attract private investments.
- Renovate educational system.
- Develop the digital economy ecosystem.
- Implement projects involving the public in digitalization, increasing their awareness and interest in digital services.
- Legitimize digital services.
- Systematize governmental supportive policies.
- Create standards defining the digital economy, its elements, and principles.
- Build an expert community.

CONCLUSION

Digital economy fundamentally changed the world by the innovative development prospects. These changes are ubiquitous: they affect individual lifestyles and change the national economy landscapes leaving behind the outsiders who were late to adopt the digital technologies. Although Russia used to be among these outsiders, it still has a potential required to catch up with the leaders and compete.

Our analysis of the Russian digital economy demonstrated that despite the constraining factors, the country can boost its digital potential to achieve the desired goal of tripling its digital economy size. The experience of other countries revealed that the further growth is not so intense, and the number of leaders may increase. The trend for digital economy development is obvious as further lag generates critical dependence, reduces business competitiveness, and undermines national security and independence.

Studying Russian digitalization success factors allowed detecting the problematic areas and suggesting recommendations for businesses and Government. These guidelines offer a tradeoff-based solution to the digital transformation crisis.

2.2   Informed Digitalization: Applying the Case Method

To support the decision-making ITC framework for crisis management and sustainable innovative development, let us discuss Gill's "Informing with Case Method"

[7]. As this book reads, in 1980s when the author was an MBA student at Harvard Business School, the curriculum included nearly 900 case discussions (i.e., 15 per week, or 3 per school day). The preparation for each discussion was very intensive and time consuming as it required deep diving into a 20-page long story, around half of which being a set of "exhibits". These typically included magazine and newspaper clippings, charts, diagrams, photos, etc. Surprisingly, these facts demonstrate that this case method was treated by the top US business school as of extreme importance. The other student activities were very limited; however, this case-based training plan still resulted in a great educational success. Why was that so?

The idea was to form an attitude, which relates to higher levels of Bloom's taxonomy, in order to make a well-reasoned decision, or at least a position to defend student's opinion. In the subsequent discussion, the professor acted as a facilitator rather than an administrator. This resembles Scrum and other agile techniques for developing software in uncertainty (i.e., crisis), where the management is typically informal, and the team is usually self-consistent and self-manageable. Also, this facilitation-based approach is closer related to mentoring than to formal management. The same approach is applied in the tradeoff-based methods such as ACDM/ATAM, which are specifically designed to reduce uncertainty, i.e., to manage crises in software development.

The subtitle of the book says: "A Guide to Case Method Research, Writing, & Facilitation". This means that the method is applicable not only to learning as generally understood and intended but also for research which is the primary focus of this book. For Software Engineering as the problem domain, the case method provides a great value of subtle requirement elicitation as it illustrates uncertain aspects in a clearly comprehensible way provided the deep-dive attitude of the software analysts/architects has been formed.

The approach to instructing through case method should be delicate. In fact, as Gill argues (also following Rangan's approach [18]), the instructor should avoid at least these things being "quasi-case" practices:

- Lecturing a case.
- Theorizing a case.
- Illustrating a case.
- Choreographing a case.
- Dialoguing a case.

As instructing by means of case method is generally out of scope of this book, we are not going to further elaborate on this subject. However, in this view, the same case method should also be applied to research with extremely great care as it can otherwise generate wrong hypotheses resulting in wrong research outcomes. This is even more potentially dangerous if this method is being applied to large-scale and consequently complex (and intricate) systems as their complexity is often a source of crisis by itself.

Avoiding these dangerous pitfalls paves way towards making informed decisions on system development. Therefore, case-based research provides helpful guidelines for successful, crisis-resistant development even for complex systems. Applying this

Fig. 2.5 A basic oscillating circuit

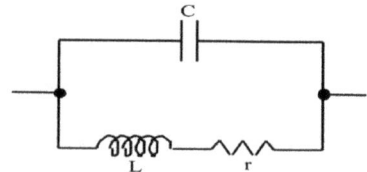

case method harnesses the human factors that we cover in more detail in Chap. 6, and in the book entirely dedicated to these mission-critical factors [28].

The approach we follow is based on the IT Crisisology framework [29]. This framework involves a few informing science aspects as its essential ingredients. Under this approach, a communication system is modelled as a three-component structure which includes:

- Transmitter, i.e., initiator of the message in the communication.
- Receiver, i.e., recipient of the message in the communication.
- Channel, i.e., communication environment.

This model is based on the so-called information theory and is representable by a circuit, which consists of capacitor and inductive coupling; it uses feedback to oscillate (Fig. 2.5) [25]. Further, while travelling in the channel from the transmitter to the receiver, the information (i.e., message) may get distorted due to an effect typically referred to as "noise". To compensate the negative effects of such a "noisy" channel, they generally recommend adding a two-way feedback loop (i.e., positive and negative). The other way to improve communication is by adding an amplifier that increases the signal value in the channel between the transmitter and the receiver (Fig. 2.6).

The case method, if applied carefully, provides feedback similar to this compensator; it also helps to amplify the signal. In this view, case method promotes resonant communication between the information exchange sides, and therefore decreases information loss and facilitates better knowledge transfer quality. In case of large-scale systems, this is mission-critical as each of the two parts, the transmitter, and the receiver, is typically a complex system in itself. The case method, thereby, serves as a source of helpful resonance in the communication channel. *Resonance* is often harmful as it typically increases signal amplitude in an uncontrolled way, which may damage and even destroy the communication channel. However, if applied carefully, resonance may improve communication due to signal amplitude increase; to do this in

Fig. 2.6 An improved oscillating circuit with an amplifier and two-way feedback

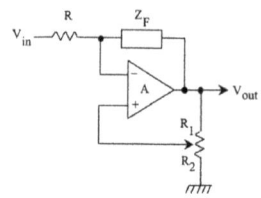

an accurate and predictable way, the mechanism should feature the above-mentioned compensation feedback loops.

Moreover, in view of system complexity, to master the reasoning and decision-making abilities under uncertain (i.e., crisis provoking) conditions the right attitude for a student (as a future expert researcher or development team leader) could be brought up by a skilled facilitator only. This skilled facilitator should be very well informed of each student (expert, or researcher) background, their personal strengths and weaknesses, conducting their discussion by minor and extremely precise actions and directing the intermediate steps so that they eventually lead the audience (i.e., the communicating system) to a new maturity level.

This looks very similar to the initial information channel model, and its enhancements are added to compensate and harmonize the potential crisis-like effect of a harmful resonance. The multi-context nature of a thoroughly designed case study multiplied by the individual diversity (in background, maturity, language, perception, etc.) of each communicating side (both of which are teams) makes the complexity, which indicates system potential to survive in a crisis.

In his book, Gill addresses case study composition and narration. To provide a multi-faceted view, he gives three different interpretations of a plot. To avoid time waste and possible confusions of the reader, Gill centres his case studies around a story known by the majority since their childhood. This is a famous fairy tale concerning the three little pigs who built their own houses to protect themselves from a dangerous wolf.

However, in contrast to the original story, there are a couple of aspects worth mentioning. First, the case is related to decision-making by means of balancing a number of factors. The IT Crisisology framework groups these into technological, business, and human-related ones. Second, the task that the little pigs are going to solve is in fact a complex one as it requires multi-criteria optimization. The story runs that each of the main characters has got a limited time and budget. Therefore, each of these activities, in fact is a project.

Moreover, within these time and budget limits they must complete a number of mutually contradictory tasks such as building a house, buying a vehicle, purchasing a set of furniture, and planning a vacation. This means that each of them has several simultaneous projects, and this concurrency essentially adds complexity. Further, another challenge of their deliverables is the location uncertainty. Many other aspects are mutually dependent; these include construction materials, furniture parameters, and security options. This is very similar to managing software engineering and system development, which is comparable to erecting a building or constructing a bridge. In these projects, several concurrent teams perform interdependent tasks such as designing disaster-proof, secure and safe environment, constructing the parts of the deliverable (i.e., foundation, load-bearing walls, sidewalks, parking lots, etc.), establishing various network-based communications (such as electric, wi-fi, ventilation, and plumbing), and so on.

At the early stage of high-level design, one mission-critical point is discovering the essential quality attributes (QA), such as security, safety, user/environment friendliness, reliability, and extendibility, to name a few. Identifying, prioritizing, and incorporating these into the high-level project plan is vital, as the early-stage mistakes or design flaws often result in a crisis of product being supplied late, over budget, or even totally undeliverable.

Each of these concurrent and mutually dependent projects requires prioritizing goals and analysing alternative scenarios. In the perspective of software engineering, these are skills required for the entire product development lifecycle such as high-level planning, architecting, implementing, testing, and maintaining the system. However, this approach may address not only technical but also more abstract and high-level skills such as analysis, decision-making, and strategic management.

Concerning optimization, this is not applied in a mathematically strict terms as a search of minimum/maximum value of a continuous function. Instead, this is a search for a good enough solution (i.e., balanced in view of the decision-maker) in a discrete space of multiple dimensions of the criteria. Therefore, this approach is searching for a compromise, which probably is not the best in terms of each individual criterion, and generally is not ideally optimal in the strict sense of multiple-aspect optimization. However, this is an operational solution applicable to the real world in a realistic time of decision-making, planning, designing, and implementing, which would satisfy the customer's requirements in terms of business, technology, and human factors. Again, this case plot looks very helpful for training in decision-making for both business management in general and software engineering in particular.

Coming back to the story itself, the characters are referred to as Alan, Brian, and Charlene starting from A, B, and C for the sake of usability. Unlike the original story, one of the three little pigs is a girl, whereas the two others are boys. The girl as a character is introduced perhaps to emphasize the way decision is made as the boys demonstrate a clearly suboptimal approach. In Alan's and Brian's decision-making, secondary aspects (such as vehicle or furniture) obviously dominate, although the case study suggests giving top priority to security in disaster-resistant facilities (such as electric hurricane window shutter) and the house itself. Also, for more straightforward and negligent decision-makers, the "shinier", i.e., more prestigious, optimization attributes seem predominant. These aspects resulted in buying a larger house (e.g., 5,000 sq. ft. for Alan vs. 1,000 sq. ft. for Charlene) and Alan's spending around 25% of the tight budget for the "mud"-Jacuzzi and luxury furniture vs. 8% of that for the simple furnishing of Charlene's house.

Other notable aspects of case design essential for the ITC framework include:

- Addressing multiple contexts.
- Scenario-based approach.
- Uncertainty.
- No single-best solution (i.e., a "silver bullet", or a solution that fits any purpose).
- Empirical approach rather than mathematically rigorous optimization.
- Addressing QA through both qualitative and quantitative assessment.

Fig. 2.7 Case study exhibit: a handwritten map (*source* [7])

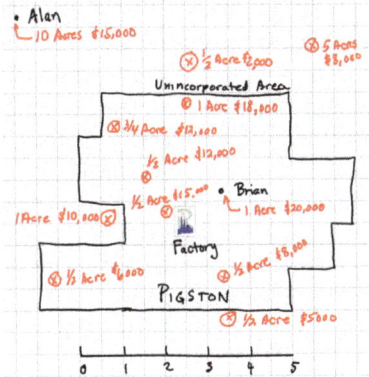

Gill introduces three versions of this case study for better understanding case design depending on the purpose (such as training and research). These differ in the following aspects:

- Structure.
- Narration flow.
- Exhibits.

Gill gives a detailed description of these aspects and their relations to the case study purposes in [7]. To make the plot more realistic and provide more food for thought, he includes such exhibits as "newspaper clips" and handwritten plans, charts, and diagrams (Fig. 2.7 presents one instance of these).

These "insights" add value to the case study, whether applied to instruction or research.

Pre-Conclusion

This chapter outlined approaches to digital transformation focusing on the Russian society in general and the country's economy in particular. First, we defined the concept of digitalization. Further, we discussed its application to the context of Russian smart and innovative initiatives in view of Industry 4.0 technologies. This discussion was based on a comprehensive overview of the historical background extracted from a collection of surveys by world leading experts in innovation and sustainable development including Ernst & Young, KPMG, Pricewaterhouse-Coopers, McKinsey, and Accenture. The purpose of this historical overview was to identify the potential impact of the new and emerging digital trends on the society in order to achieve competitive benefits and ensure sustainable development of the national economy.

Based on this survey, we identified the points of potential growth, one of which was IT-intensive services located on top of internationally or locally recognized digital platforms such as Yandex.Market, VKontakte, Facebook, Amazon, Airbnb, and a few others.

Next, we analysed the factors that trigger successful digitalization. These help avoiding dangerous pitfalls, providing sustainable development, improving production quality, introducing mass-scale innovations, etc. At the national level, these strategies include the National Technology Initiative, which is a long-term program promoting the leadership of Russian companies in new and emerging high-tech markets based on blockchain, IoT, and a few similar advanced technologies. The domains of particular importance include intellectual control systems, personalized health care, clean energy, self-driving vehicles, cyber-physical systems, and more.

Afterwards, we identified the basis for the above domains that included the following technologies:

- Big data.
- Neurotechnology and artificial intelligence.
- Distributed registry systems (i.e., blockchain).
- Quantum-based technologies.
- Industrial Internet.
- Robotics, wireless, and sensor components.
- Virtual and augmented reality.

Based on the above considerations and smart business cases (such as Tinkoff Bank, Yandex, and a few other instances), we produced a set of recommendations for successful digitalization and sustainable development.

Finally, we addressed the digital transformation following the IT Crisisology framework introduced in our earlier books (e.g., [28, 29]). In this view, such transformation is typically equivalent to monitoring, addressing, and managing a crisis, which requires decision-making under uncertainty, and thorough balancing of business, technological, and human-related factors. To do this in a smart way, we suggested using a case study as a powerful research method, which is based on informing science approach, and provided an example of this method application as a concept proof.

2.3　Case Walkthrough: Old Fairy Tale Revisited

Gill refers to a walkthrough as "… a complex example that typically requires the facilitator to lead participants through a series of steps" [7]. Before the final conclusion, let us focus on it in more detail as we believe this is very important for a case study.

Recollecting the plot of Gill's sample case study, its main characters are named Alan, Brian, and Charlene (starting from A, B, and C to concentrate on the analysis rather than the names). Different from the original story, one of the siblings in this case study is a female, and the two others are males. Gill introduces the girl probably to highlight the differences in decision-making by experts of different proficiency levels as the boys perform in a suboptimal way. For Alan and Brian, decision-making has a single dominating factor out of suboptimal ones (e.g., vacation or vehicle), even though the case study suggests security as a mission-critical aspect to resist/avoid the disaster (such as hurricane protected windows and the general house architecture). Additionally, less experienced, and outward decision-makers tend to prefer more prestigious factors. For instance, such aspects may include spending the tight budget for a larger house or luxury options, such as "mud-Jacuzzi" and furniture.

Let us revisit each of the three versions of the famous tale separately, and try to identify interdependencies, including primary logical links and cause-and-effect associations.

Each case study comprises a 3-page text and a supplemental appendix section featuring a more detailed "attachments" aimed to provide the reader (either a student or a researcher) with a meaningful insight. The first two instances have a brief single-page supplemental section of just one table, whereas the final case incorporates as many as five "exhibits" 7 pages long, which include a handwritten map of the local area, a few tables, and even a 2-page newspaper clip.

The first story (referred to as Version A in the book) starts in a reverse order, as it presents the narrow escape of Charlene's brothers from the wolf's attack, and the criminal being imprisoned. The introductory section immediately takes the reader into an urgent real estate sale of the girl's tragically passed away parents, which resulted in a $300,000 net income. After this brisk introduction, and a very quick one-paragraph background, the next subheading reads "The Decision". Clearly, this case study is about decision-making.

The income received by the relatives should be equally shared between the young lady (yet being the eldest in the family and, therefore, responsible for the deal) and her two brothers, Alan and Brian. As the eldest, she was responsible for the presale preparations of the property. The story tells us that the younger brothers, overwhelmed with this sudden good news, made relatively careless decisions. Before the deal was closed, they managed not only to get their shares, but also to spend them. Obviously, such abrupt actions were immature and suboptimal in terms of decision-making. Conversely, the girl acted in a much better prepared, no-rush, and well-justified way. Apart from the experience resulting from her being older, the reason for Charlene's maturity as an expert was her well-balanced understanding of personal asset management policy. She understood that this relatively large asset, although received easily, should be spent wisely to serve as a long-term insurance for disaster coverage. Perhaps, she noticed the upcoming crisis signs as a wolf was going to visit a mostly careless and short-focused Pigston community.

"The Decision" section includes repeated evidence based on multiple criteria, e.g., "A variety of factors had weighed into the decision", or "A number of factors had come into play". The task of balancing within this variety of alternative solutions required a thorough analysis, hard to perform in a brisk and straightforward manner. Some of these factors to be balanced included estate cost, distance from home to work, and transportation availability. As the brothers performed a quicker, surface-level analysis, they were unable to incorporate the entire set of all these contradictory factors. Charlene, as a more advanced analyst, wisely attempted to embrace as many essential factors as possible. She stated that "… sacrifices had to be made. In the long run, I felt I'd be better off with a nice house in a great neighbourhood that I could sell at a profit some years down the road".

Gradually, more factors were introduced into the story, one of them being the real estate safety. The dangerous wolves wandering around threatened to destroy any lightweight building constructed of straw or similar materials. Even relatively thin brick walls, which seemed "wolf-proof" at the first sight, were indeed at a serious risk. At that time, the boys revisited their early strategies based on surface factor analysis and began considering these new threats and some other important factors; however, it was too late as their incomes had already been invested. The final section, named "Lessons Learned", reconsiders these factors and gives more food for a balanced decision-making. Next, the case lists a few sections of exhibits which bring together the priorities that appear in the text of the story. The factors listed are the expenses, including:

- Housing.
- Lot.
- Furnishings.
- Transportation.
- Travel.
- Mortgage.

The second case study (referred to as Version B in the book) is framed similarly to the previous one, although its narration is not so straightforward; perhaps, this is intended for a more advanced reader. The third case study looks even more advanced; its exhibit collection is the longest and the most attractive; it exceeds each of the two preceding stories by over 3 times in length and features 5 times more exhibits that include a colour illustration looking as if being hand-drawn, and even a supplement referred to as "Article from Pigston Pigayune".

The second case study looks more decorated than the first one; it starts with a popular song titled "Who's afraid of the big bad wolf?" and recollected by Charlene. This song deals with the reasons to be afraid of the wolf, which has just attacked her brothers, and is currently threatening herself. Immediately, and similarly to the previous instance, the background follows. This background section contains the overall financial amount to be shared between the siblings, and the other preconditions important for the case. This section concludes that Charlene should rethink before deciding on her choice as such good fortune happens seldom in one's life.

The next section specifically deals with the house selection, drawing attention to prioritizing the criteria available (such as furnishing, transportation etc. discussed further). The text explicitly says that she spent $95,000, i.e., 95% of the budget for this critically important option due to the mission-critical nature of security QA: "Citing security as her top priority, Charlene had opted for every construction safety feature available". The variety of options and choices of the three siblings are given in Exhibit 1 framed exactly as the previous instance.

Further, the case discusses the threats originating from the dangerous wolf, who can break into houses made of straw and even bricks featuring no special protection. This additional data says that although the options available suggest a rich variety of alternatives, the actual expert choice is more constrained. Again, this is explicit in the text: "Naturally, that choice could not be made independently of housing". One more reason for that housing-dependent selection is the extremely high value of the property in the proximity of the factory, the major local employer. In this respect, the property within a short walking distance from the factory is unaffordable for any of the siblings, and they have to consider buying a means of transportation to keep safe while travelling. This imposes extra constraint on their budgets and calls for rethinking of their decisions.

The next section, "The Big Bad Wolf", describes the wolf attacking one of the brothers, Brian, and momentarily destroying his straw house. In this attack, the boy suddenly remembered the party when the two brothers sang the famous "Who's afraid of the big bad wolf?" hit and laughed out loud as they did not understand the reality of the threatening attack. This song's refrain of "…singing the then-popular "Who's afraid…?" is a reminder for the reader to reconsider the possible scenarios and re-prioritize the options according to the re-sorted criteria. The neighbouring Alan's lightweight house constructed similarly to Brian's was also ruined momentarily. Importantly, the wolf's attack destroyed not only the boys' houses but also their luxury cars. These were Alan's "P150 pickup truck with chrome detailing" and Brian's "Porkhe sports car" also referred to as "sow magnet" and "German sports car". These cars, being "shiny" objects made the boys spend $35,000 and $40,000, i.e., 35% and 40% of the total budget respectively. Still, they both appeared to be poor investments as they were totally ineffective in the sudden and harsh wolf's attack.

The final section is devoted to the decision-making process at the moment the wolf attacked Charlene's house. Again, there was a number of scenarios: run to a better protected place such as a market or police station, distract the wolf with foodstuffs, and some other options. However, the relatives remained safe as the house was strong enough to withstand the invasion.

Looking at Exhibit 1 (i.e., budget breakdown table) with more attention, we can conclude that the total investments for the boys were substantially higher than that of Charlene (15% for Alan and 40% for Brian respectively). This resulted in heavy mortgage burdens for the boys, whereas Charlene stayed with a zero mortgage as the exhibit reports. Another observation is that based on the over-budgeting, the boys could afford a choice of a better housing at the cost similar to the girl's house (the respective amounts were $30,000, $40,000, and $50,000). Again, their preferences were very different as Alan paid for luxury options (including Jacuzzi and furnishings), whereas Brian invested in a spacious 1600sq. ft house. The housing materials were also very different: straw for Alan, stick frame for Brian, and bricks for the girl. Interestingly, the basic housing and furnishing costs (excluding Charlene's security options) were approximately the same for each of the three characters: $50,000, $50,000, and $54,000 respectively. Still, their investment breakdowns were radically different as we discussed, the reason being maturity level.

Another factor the siblings considered was the "lot", i.e., the cost of the land for the house, which depended on both position and area. Again, priorities were critically different. The first boy chose a huge and relatively cheap $15,000 ten-acre space in a dangerous and remote Pigston city suburb. The second boy spent $40,000 for a luxurious and very expensive golf course one-acre lot in a safe location, and the girl bought a twice smaller lot at the same location. This single one-page exhibit appears valuable, as we can see that the boys' selections were not totally unreasonable but also relatively balanced, although their focuses, i.e., decision-making criteria were misplaced. In software engineering, this is known as "scope creep"; let us reserve this observation for a later discussion.

The third case study (referred to as Version C in the book) as we already noticed is much more creative in terms of text style and has more detailed exhibits. This case, perhaps, follows Gill's intention to demonstrate the difference between a "general purpose" case study and that for a more advanced reader. Surprisingly, this intention does not affect the text size. However, details, i.e., exhibits matter. Conversely to previous cases with 75% of text and 25% of supplementary materials, Version C contains an inverted, i.e., "flipped" layout: the exhibits dominate in size and make 70% of the total case volume, whereas the text amounts to 30% only. Moreover, the appendices are multiple as opposed to the previous versions, and different in size and shape. They include a hand-drawn map of the Pigston city area and surroundings, a newspaper article clip, and a richer set of tables detailing the selection options for the decision-makers. Let us discuss each of the ingredients of this sophisticated example in more detail.

The hand-drawn diagram is a map authored by Charlene who is a trusted information source. The map presents certain lots in the Pigston city area. Explicitly, the map gives an idea of the property sizes and costs. Implicitly, it also outlines the distances between the assets and the factory suggesting transportation options, including walkable opportunities. The exhibits are linked to each other; for instance, the text mentions the "unincorporated area" where Alan bought his property, and the map marks both locations.

Exhibits 2 and 3 present the tables for transportation modes and building specifications, respectively, together with their costs. These provide more details to make a better justified choice. These tables even feature columns for comments, which sometimes are distracting (e.g., "Owners should not use "wee, wee, wee" as entry phrase"), and sometimes meaningful (e.g., "Primarily for aesthetics and noise reduction", "Foreign models, such as the Porkhe, tend to run $5,000-$10,000 higher than domestic models", etc.). This helps avoiding over-budgeting and concentrating on mission-critical criteria such as security and safety.

Exhibit 4 is similar to the comparative budget breakdown tables of the previous cases; strange as it may seem, it contains the boys' expenses only. Perhaps, the idea is the same as in the previous tables for this case study: to demonstrate extra costs and suggest saving options implicitly. Again, this case sample suggests design ideas for advanced readers. Examples of the extra expenses include:

- 5000 sq. ft straw house ($25,000, Alan).
- Mud Jacuzzi ($5,000, Alan).
- 1600 sq. ft stick-frame house ($40,000, Brian).
- 10-acre mini estate in unincorporated Pigston exurbs ($15,000, Alan).
- 1-acre double golf course lot in Pigston area ($40,000, Brian).

- "Party Hearty" complete set from William Sownoma ($25,000, Alan).
- 5 rooms of exotic furnishings from Pig 1 ($10,000, Brian).
- P150 pickup truck with chrome detailing ($35,000, Alan).
- Porkhe sports car ($40,000, Brian).
- 3-week vacation in Europe ($10,000, Alan and Brian each).

The final Exhibit 5 probably refers to Gill's student experience at Harvard Business School, heavily based on case studies. This is a two-page excerpt from a "newspaper article" providing "evidence" for the case. This contains a warning quotation by Pigston's Mayor: "The best thing to do in the extremely unlikely event that a rogue wolf is sighted is to hide. Once a rogue wolf sees you, it will follow you to the ends of the earth, using its highly tuned sense of smell—the most sensitive in the sentient animal community. Also, forget about hiding in your house. Straw, wood, even many brick dwellings can't stand up to the 175 mile an hour wind that a bellows wolf can produce". These words present wolf's attack as a critical yet extremely rare event. Clearly, this can be marked as a potential crisis with a possibility yet to be determined. Although the Mayor tries to persuade the observer that the chance of such an emergency is extremely low, this probably should be addressed as a potentially critical event.

After presenting the walkthroughs of the three case study versions, let us summarize their similarities and points of difference. From there, we will be able to ascertain their common dependencies and the anticipated consequences and discuss their possible applications within the framework of IT Crisisology.

Naturally, the basic layouts of the story, well known to many since early childhood, are quite similar. This is an evident benefit as the plot does not distract the reader's attention to comprehend it; instead, we may focus on more specific aspects. What really matters is the complexity artificially added to this well-known story; this is central to either realistic or real-life situations that case studies typically describe. This complexity has at least two dimensions: multiple aspects or "factors" as Gill calls them, and a number of constraints that complicate the decision-making process. Similar to real life, these constraints apply to resources, such as time and budget. The budget is limited by a moderate amount of $100,000 which should be balanced between a few heterogeneous objects. These include a piece of land, a house, a vehicle, and a number of other items. Time is limited as the dangerous wolf is approaching the "greater Pigston area" threatening its unprepared inhabitants. As in real life, it is possible to go overtime or over budget; however, these are risky and expensive options available to aggressive investors and brutal decision-makers.

The two brothers, Alan and Brian, are given as an example of this kind of decision-makers, in case we do not call them careless. This over-budgeting made them face a crisis as they lost their properties, i.e., both houses and cars, at the same time having to pay the mortgage. Obviously, their limited ability to analyse concurrent factors made their selections suboptimal.

The other ability this case study trains is prioritizing multiple criteria. The boys tend to prefer visually attractive factors, whereas the girl performs a more justified analysis by prioritizing her choices. Security is her top priority; choosing a number of expensive options to guard her living space, such as electric hurricane window shutters and tile roof with hurricane straps, she sacrifices the comfort and luxury options due to budget constraints.

We mentioned optimization several times. Indeed, the problem that every version of this case study addresses is multi-criteria optimization. In contrast to traditional mathematically sound optimization of a smooth function, this kind of optimization suggests a discrete number of options to select from. This number is relatively large, so the choice is not trivial, although the plot may look simple. This kind of optimization is close to the competence of justified decision-making, which is essential for any software engineer. Therefore, we consider the case study approach as a required ingredient in both student training process and IT Crisisology framework.

The key difference points of the case study versions are:

- Level of detail.
- Narration flow complexity (i.e., straightforward or more sophisticated).
- Number of artifacts attached as exhibits.

Coming back to the crisis, we also encountered this issue in our discussion.

Let us make a conclusion regarding the three versions of case study design and layout as presented above.

Our discussion centres around decision-making, which as an ability is a critical skill in many professions and for many competencies including software engineering and crisis management. This decision-making as a process includes the following sub-processes:

(i)   Selecting criteria.
(ii)   Forming/choosing a set of alternatives.
(iii) Analysing these alternatives according to the criteria.
(iv)   Selecting the best alternative in terms of the criteria set.

At this point, let us note a few important details. First, in our case, the decision is based on multiple criteria. Second, these criteria should be prioritized. Third, the selection made is not necessarily ideal (i.e., the best); rather, it is good enough to satisfy all the critically important criteria in a balanced way.

This decision-making is the crisis-related process, due to the following factors:

(i)   Complexity.
(ii)   Uncertainty.
(iii) Human factors such as gender differences in decision-making.
(iv)   Maturity level: Charlene behaves more as an expert, whereas the boys' decisions are less mature.

The other crisis signs include:

(i)   Time pressure: as the wolf is approaching, it is impossible to analyse/implement every potential alternative.
(ii)   Tight budget: going over budget is possible; however, such a decision results in an additional financial burden, i.e., mortgage.
(iii)   Unexpected factors: the wolf destroyed not only the shelter but also the means of transportation, which required immediate response as it critically influenced multiple resources, e.g., time and budget.

The decision-making process has a certain lifecycle, which resembles the PDCA/DMAIC approach and includes the following steps:

(i)   Formulate the goal.
(ii)   Detect the important factors.
(iii)   Analyse the important factors.
(iv)   Select the critical factors.
(v)   Prioritize the critical factors.
(vi)   Breakdown the resources available.
(vii)   Validate the decision.
(viii) (Re)iterate if/while necessary.

At this point, it is worth noting a few important aspects. First, Step (vi) of the process should be balanced in terms of budget, time, and labour. Second, Step (vii) of the process should include a validity check for the goal to be realistic/achievable, e.g., according to well-known SMART principles. Third, the lifecycle may (and typically does) require several iterations, which in turn involve a new round of selecting, prioritizing, validating, etc. before finalizing the decision-making process.

The case study was presented in three different versions intended for different levels of judgement. Version (A) was aimed at the basic level of judgement as it featured straightforward narration. Version (B) was aimed at the advanced level of judgement as it included certain subtle and/or implicit fragments, such as prioritizing criteria and pointing out security as a possible top priority. Version (C) was aimed at the expert level of judgement as it comprised multiple and more detailed exhibits such as hand-drawn map, and newspaper clip.

To apply these case studies to real-world practice of decision-making, specifically in software engineering, let us name the applicable methods for their implementation:

- Discrete optimization methods including ACDM/ATAM for the initial solution design/layout.
- Analysis methods including AHP.
- Agile methods including Scrum, XP, and OUP.
- Soft skills including teamwork, critical thinking, and negotiation.

The real-world applications require an even harder choice in terms of resources including time pressure, budget constraints, and requirement changes, i.e., "scope creep", and therefore reconsidering the priorities and applying continuous iterative adjustment (e.g., based on PDCA/DMAIC lifecycle).

The Little Pigs Stories demonstrate that brute force approach is inapplicable in most real-life cases; the decision-making (and specifically analysis) process requires thorough optimization. In real-life solutions, careful selection of the options available/applicable can downsize the decision range significantly (at least by 30% according to our estimates).

To conclude, let us summarize a recent interview by a famous HSE professor and former Deputy Head on Research, Lev Lyubimov (1936–2021) [11].

Currently, it is very easy to get information: the internet as its global source presents an incredible volume of data. We no longer need visiting libraries, digging the files or waiting for a book to arrive. Flipping the pages in the gadget is very quick. With multiple history textbooks and a multitude of opinions, every reader decrypts the text according to his/her own code. At school, a student uses knowledge to instantly solve the problem. In this context, we should teach the student the way of analytical reading rather than merely reading. A rare primary school teacher knows what reading really is. Reading is a complex communicative behaviour that extracts the meaning. As behaviour is central, the student has to join the story and become an active participant, a character of that text. This is a mission-critical skill to teach.

Shifting the focus from a passive bystander or a spectator/onlooker to an actor, i.e., an active participant of the story is essential for crisis responsive decision-making. Case studies provide an invaluable opportunity for training this skill of the active reader's involvement, no matter being a young student or an expert researcher. The reader must be an active participant of the plot; the case study is specifically intended to make him/her do this. Transforming the reader into an actor produces a dramatic change in perception and hands-on learning of the new concept application to smart and innovative technologies and provides sustainable professional development in such complex areas as computer science and software engineering.

2.4   Conclusion

This Chapter presented the phenomenon of organizational digital transformation i.e., digitalization. First, we defined the key terms, strategic ideas, and important principles and examined this complex and challenging phenomenon based on the framework of IT Crisisology. In this view, our primary research method was case studies. Specifically, we applied the case-based reasoning to identifying success factors of the Russian economy in the undergoing process of its digitalization.

Next, we analysed the structural changes that digitalization already brought to the world, and the emerging development opportunities. These changes appeared mission-critical as they dramatically affect individuals, businesses, national, and global economy. The IT Crisisology assisted in informed decision-making in the digitalization crisis as Russia appeared to have a certain lag behind the digital leaders such as US, and certain EU states. However, the country has enough potential to get competitive; therefore, despite the constraining factors detected, our analysis revealed that there is a way to boost the Russian economy by digitalization. Yet, this catch up should be started immediately as the further pace of the leaders tends slowing down. Otherwise, any further delays might result in a national crisis that would not only weaken competitiveness but also critically affect sustainable development of the country. Thereby, ITC appeared a practically applicable, multi-faceted approach to examine such a complex phenomenon as digital transformation and validate the concept proof.

References

1. Annual reports of PJSC MegaFon [E-resource]. PJSC MegaFon. Retrieved from https://corp. megafon.ru/investoram/shareholder/year_report/ (Last Accessed February 17, 2022).
2. Butenko, V. Starring: Internet.ru—the economy of new opportunities. Vedomosti. Forum application.
3. «Digital Russia: a new reality» [E-resource]. McKinsey Retrieved from https://www.mck insey.com/~/media/McKinsey/Locations/Europe%20and%20Middle%20East/Russia/Our% 20Insights/Digital%20Russia/Digital-Russia-report.ashx (Last Accessed February 17, 2022).
4. Dombrovsky, Y. (n.d). Technology and people [E-resource]. Retrieved from Retrieved from http://www.bfm.ru/news/181410 (Last Accessed February 17, 2022).
5. "Dmitry Medvedev has approved a roadmap to Russia's digital future"/[E-resource]. Rossiyskaya Gazeta. Retrieved from https://rg.ru/2017/07/31/medvedev-utverdil-plan-razvit iia-cifrovoj-ekonomiki.html (Last Accessed February 17, 2022).
6. «Evangelist». [E-resource]. Journal The Village. Retrieved from http://www.the-village.ru/vil lage/business/newprof/146891-evangelist (Last Accessed February 17, 2022).
7. Gill, T. G. (2011). *Informing with the Case Method: A guide to Case Method Research, Writing, & Facilitation* (p. 563). Informing Science Press.
8. IFR Press Conference presentation//[E-resource]/International Federation of Robotics.URL: Retrieved from https://ifr.org/img/uploads/presentation_market_overviewworld_robo tics_29_9_2016.pdf (Last Accessed February 17, 2022).
9. Industry 4.0: Building the digital enterprise [E-resource]. Price WaterhouseCoopers. Retrieved from https://www.pwc.ru/en/technology/assets/global_industry-2016_eng.pdf (Last Accessed February 17, 2022).
10. Kozlova, K. A., & Markova, E. G. "Venture financing of innovative projects".
11. Lyubimov, L. (2018). Believing in service and mission. *News Companion Journal*. Retrieved from https://www.hse.ru/our/news/459386207.html (Last Accessed February 17, 2022).
12. Materials of the expert discussion "Problems of Forecasting and Modeling the Labor Market in Russia" [E-resource]. Scientific Bulletin of the IEP. Gaidar. Retrieved from http://iep.ru/files/ nauchniy_vestnik.ru/1-2016/40-61.pdf (Last Accessed February 17, 2022)
13. More than 15 billion rubles will be invested in the development of machine tool building by 2016. [E-resource]. Ministry of Industry and Trade of the Russian Federation. Retrieved from http://minpromtorg.gov.ru/press-centre/news/#!8640. (Last Accessed February 17, 2022).
14. New South Wales: Unified Public Services. [E-resource] Open Government. Retrieved from http://open.gov.ru/events/5512186/. (Last Accessed February 17, 2022).
15. Presidential Decree of May 9, (2017) No. 203 "On the Strategy for the Information Society Development in the Russian Federation for 2017–2030"
16. Putin: the formation of a digital economy is a matter of national security of the Russian Federa-tion. [E-resource] Tass News Agency. Retrieved from http://tass.ru/ekonomika/4389411 (Last Accessed February 17, 2022).
17. "PwC Global Digital IQ Survey 2015" [E-resource]/Pricewaterhousecoopers URL: https:// www.pwc.com/gx/en/industries/healthcare/publications/digital-divide.html (Last Accessed February 17, 2022).

18. Rangan, K. (1995). Choreographing a case class, HBS Case Note, 9-595-074.
19. Regulation of the Digital Economy. [E-resource]. Retrieved from http://www.tadviser.ru/index. php/%D0%A1%D1%82%D0%B0%D1%82%D1%8C%D1%8F:%D0%9D%D0%BE%D1% 80%D0%BC%D0%B0%D1%82%D0%B8%D0%B2%D0%BD%D0%BE%D0%B5_%D1% 80%D0%B5%D0%B3%D1%83%D0%BB%D0%B8%D1%80%D0%BE%D0%B2%D0% B0%D0%BD%D0%B8%D0%B5_%D0%A6%D0%B8%D1%84%D1%80%D0%BE%D0% B2%D0%BE%D0%B9_%D1%8D%D0%BA%D0%BE%D0%BD%D0%BE%D0%BC% D0%B8%D0%BA%D0%B8 (Last Accessed February 17, 2022).
20. Selishcheva, T. A. (2021). Problems of information inequality of the Russian economic space. *Scientific Journal VESTNIK*. St. Petersburg State University c. 4–12.
21. Shannon, K. (1963). *Works on the theory of information and cybernetics* (p. 832). Foreign Literature Publishing House.
22. «Sibur»: digital-revolution in oil-and-gas industry. [E-resource]. Retrieved from https://hh.ru/ article/312408 (Last Accessed February 17, 2022).
23. SME banking in Europe. [E-resource]. Retrieved from Finalta https://cis.smebanking.club/ store/sme-banking-europe/ (Last Accessed February 17, 2022).
24. The Ministry of Telecom and Mass Communications of Russia has presented a rating of regions in terms of the level of development of the information society. [E-resource]. Retrieved from Ministry of Telecom and Mass Communications of Russia http://minsvyaz.ru/ru/events/35027/ (Last Accessed February 17, 2022).
25. Tinkoff Investments will be untied from BCS. [E-resource]. Retrieved from Financier, https:// finansist-kras.ru/news/analitics/tinkoff-investitsii-otvyazhut-ot-bks/ (Last Accessed February 17, 2022).
26. Vasetskaya, N. O., & Gaevskaya, T. B. (2018). Analysis of the structure and problems of the program "digital economy of the Russian Federation." In *Proceedings of the Scientific and Practical Conference with International Participation on April 2–4.*
27. Vladimirov, A. (n.d). The horrors of our cities [E-resource]. Retrieved from http://www.itogi. ru/russia/2013/31/192428.html (Last Accessed February 17, 2022).
28. Zykov, S. V., & Singh, A. (2020). *Agile enterprise engineering: Smart application of human factors. Models, methods, practices, case studies.* Springer Nature, 156 pp.
29. Zykov, S. V. (2021). *IT crisisology: Smart crisis management in software engineering models, methods, patterns, practices, case studies.* Springer Nature, 181 pp.

Chapter 3
Pre-digitalization: Earlier Cases

3.1 Rebranding: Transforming Anderson to Accenture

Accenture: The history of transformation

Background: company name, key facts

Accenture plc is a multinational professional service company. A Fortune Global 500 company, it reported revenues of $44.33 billion in 2020 and had 506,000 employees. Accenture was created by the separation of Andersen Consulting from its parent Arthur Andersen in 2001. Frank Modruson, Accenture's chief information officer and the person responsible for carrying forward the IT transformation challenge had ambitious plans to replace legacy systems of Arthur Andersen. Accenture's current clients include over 90 of the Fortune Global 100 and more than three-quarters of the Fortune Global 500.

Walkthrough: a brief history of the company development

The Arthur Andersen accounting firm was founded in 1913. Accenture, initially named Andersen Consulting, started its way as a part of the Arthur Andersen company. In the year 1954, Arthur Andersen decided to dig into the field of consulting. This sector of services steadily grew. In 1989, Arthur Andersen decided to split its business into two separate entities: Andersen Consulting, in charge of all consulting activities of the firm, and Arthur Andersen, which continued to provide traditional financial audit services. Over the following years, it became clear that the spheres of activity of the companies partially overlapped. Therefore, in 1997, Andersen Consulting began a process of arbitration that sought to separate the consulting division from the financial audit firm. It took several years, and in 2001 the firm adopted its name Accenture in a $175 million rebranding campaign and became independent.

Crisis of transformation: reasons, sources, and threats

To support Accenture's launch as a newly independent business enterprise, it was granted permission to use Andersen's technology infrastructure for 1 year. And this challenge was proudly taken.

After achieving independence, such a large organization needed an effective IT infrastructure. They had the right to use Andersen's technology infrastructure for 1 year only, a very short time limit to create an IT infrastructure of their own.
    Andersen's technology had the following problems:

- Andersen's systems were a patchwork of legacy applications.
- Obsolete software platforms, key systems, and databases could not be accessed remotely through the Internet, i.e., transformed into services.
- Due to proprietary accounting and HR management software systems, it was very complex to get an up-to-date snapshot of the whole organization's status.

Crisis management plan

After considering the possible alternatives, Accenture opted for a single-vendor approach with the hope of minimizing the total cost of ownership of its IT infrastructure. To run most of its back-end IT operations, as well as to provide basic communication and productivity applications, Accenture chose Microsoft as its strategic partner.
    Accenture also handpicked a few major providers for their hardware needs. They went with HP and other suppliers for the computers and servers, and with Cisco for all their network-related equipment. Similarly, Accenture chose SAP as their worldwide application provider for financial and HR solutions.
    Accenture decided to run its e-mail infrastructure under a managed service approach.
    One of Accenture's fundamental initiatives to reduce IT costs was outsourcing.
    To support the new financial processes, Accenture sought an advanced, web-wired enterprise technology solution, one that offered greater flexibility, centralized control, robust reporting, and an ability to integrate the finance organization with other critical corporate functions.
    To increase agility in this transformation crisis, Accenture also shifted to a "core-light" personnel strategy. By 2010, only 14% of their IT staff worked directly for the company as permanent employees, whereas the rest 86% was "borrowed" via the Accenture Global Delivery Network (GDN) and the Infrastructure Outsourcing (IO) group.

Is IT mission-critical for Accenture? Why?

The IT infrastructure always was mission-critical for the company because it allowed accurate and timely information exchange between the company offices in different countries worldwide. This always was and still is one of the most important factors that allow company to provide clients with high-level services.

Accenture crisis-resistant development: the major steps

Accenture undertook an exemplary journey in transforming its IT capabilities. As the firm's global workforce more than doubled in size from 2001 to 2008, its IT organization managed to reduce its spending per employee by 60% and cut down IT overall expenses as a percentage of net revenue by 58%, at the same time increasing the personnel satisfaction of the IT tools and services.

The new Accenture vision proposed that IT should run as a business within a business rather than a cost centre.

Previously, consultants were accustomed to personalized in-office help only. This was expensive for Accenture, and most of the time their offices were overstaffed with IT service managers. After the new system implementation, consultants learned to resolve their queries mostly through the online databases (i.e., corporate knowledge-base) and were gradually trained to call or seek personalized help for urgent or severe problems only.

To build the new IT infrastructure, three major decisions were made and brought to life. The first one was to use single platform and single instance of infrastructure for all the offices. This allowed Accenture to minimize the support costs, unify the processes, and better control the internal cash flows. The second one was to adopt a small number of product families for common tasks instead of an infinitely large number of independent solutions. This allowed to offer richer functionality without an extra budget and HR involved in developing the connectors and universal interfaces. The third change was to develop software in-house thus running IT as a separate business inside the core business. This allowed Accenture to have cutting-edge and reliable solutions, reduce the costs, and better meet employee concerns.

What worked to manage the crisis, and what did not? Why?

The cost benefits of migrating to a single-platform architecture were significant. Accenture was able to move from three distinct directory systems to a unified one, from more than 400 Novell file servers to 50 Microsoft servers, and from 440 users per e-mail server to 2,500 users per exchange server.

Most Accenture steps to manage the crisis worked. However, certain radical changes were not easy to complete, and there was a noticeable pushback from different levels of the organization, especially IT staff who would be more comfortable to keep their previous roles and duties. The change made the IT staff responsible not only for operational statistics (such as server uptime, etc.) but also for dealing with people and satisfying service levels.

Is outsourcing always a success factor?

In pursuit of further cost-cutting, Accenture chose outsourcing their IT infrastructure via the newly developed Accenture GDN and the IO group. This allowed the company to respond to their own IT needs in a cost-effective manner, i.e., by rotating their staff between the countries. It also allowed to adopt the cloud infrastructure and network solutions managed by the vendors—again lowering the costs and avoiding superfluous headaches. Bob Kress, Senior Director on Accenture IT Business Operations, said: "We could be more flexible to ramp out IT force up or down if we outsourced it. Of course, we retained the project management talent and out brightest thinkers but coding itself—we knew we could do that very well in Asia". Summing up, outsourcing became one of Accenture's success factors, although not the only one. However, there are examples showing that this is not always the case. There are many well-known cases (e.g., the Royal Bank of Scotland that outsourced their IT activities to India) when the contractors could not fulfill their obligations and implement software, which caused million-dollar expenses.

Would following a process improvement lifecycle (such as DMAIC/PDCA) help Accenture to improve business?

While adopting outsourced software, Accenture ensures its stability, and this is where PDCA lifecycle is commonly used. The company strategy is to test the new versions of software and services available and upgrade skipping the newest versions to reduce the risks of instability and still benefit from the new functions.

The above process is similar to DMAIC, which is also used in Accenture for developing new software applications. First, the business benefits and the potential return on investment (ROI) are calculated, and in case they are feasible, the new software is implemented. In the next 3 years, financial audits are performed to ensure that ROI reached the value expected.

Would following an efficient process (such as Lean/Kaizen) help Accenture to improve business?

The Lean process is consistently used inside Accenture. As Kress pointed out, "… we expected IT managers to lower the costs of the department they run by 10% every year. It was an aggressive target, but it certainly kept up creative—and kept us on our feet".

Examining the Kaizen process, one could easily see that Accenture extensively uses this, and the entire outsourcing strategy is a step fully compliant with it.

In what way should IT and business strategies be aligned?

The example of Accenture demonstrates that running IT as a business inside business is a powerful strategy. This allows to define the business strategy, search for the tools required to follow the strategy, and develop a service plan with reasonable prices and variable service levels. This ensures in-house development stability, clear business focus, and competitive pricing.

In a non-IT company, can IT be a separate business area that creates value?

As the functions range and price of solutions developed are competitive, the company could potentially re-sell their in-house products and services to external customers and gain significant profits. The variability of service levels and well-trained support team with a 24/7 call centre (due to multiple time zones) allowed Accenture to achieve this.

Identify three key risks for Accenture and suggest their mitigation strategies.

Concerning the Accenture risks during the transformation, the biggest one is probably the risk of a wrong platform choice. If the platform vendor quits, the customer will lose support and future product releases. According to the case study, Accenture chose the major partners with solid financial position and reputation (such as Microsoft, SAP, HP, and Cisco), and they all are active in business to the present day.

What are the future development strategies?

Most likely, Accenture will re-organize its market-leading capabilities into these four services: Strategy & Consulting, Interactive, Technology, and Operations. The company will manage its businesses through three geographic markets—North America, Europe, and Growth Markets—instead of operating groups. Accenture will continue to go to market by industry and expand its global programs. At the same time, Accenture will be making leadership changes and expanding its Global Management Committee to include a broader representation of leaders from their services and geographic markets.

How would you plan Accenture development to fit the new digital era?

Accenture's rotation to the novel and emerging areas (such as digitalization, cloud services, and cybersecurity)—over the recent few years has demonstrated an unmatched ability to identify and scale new opportunities in the most strategic, high-growth areas of the market. These changes are designed to increase the company's ability to anticipate client needs and market changes.

3.2 Getting Going: Cirque Du Soleil

Cirque du Soleil: THE STORY OF TRANSFORMATION

Background: key facts and figures

Cirque du Soleil is an international entertainment company and the biggest circus business in the world, originally founded by Guy Laliberté in 1983. Since then, the company has grown significantly, becoming de facto, the world's most well-known circus [5].

Key figures and facts:

- Headquarters: Montreal, Quebec, Canada
- Founded: 16 June 1984
- Key person: Daniel Lamarre, President and CEO
- Revenue: $895.5 M as of 2018
- Number of employees: 1,400
- Debt: Over $1B

Walkthrough

Cirque du Soleil started as a troupe from Quebec, which performed, in various forms, from 1979 to 1983. The first official success was their Le Grand Tour du Cirque du Soleil, which caught people's attention and secured funding. After that, the business was recreated as a proper circus and had an impressive growth [4]. By 1990, it became profitable, 10 years after it had over 19 shows in some 300 cities on all the six continents. In 2017, the company generated approximately USD 1 billion in revenue.

Crisis of transformation: reason, source, and threats

Although the circus as a business at the first glance might not seem like something that needs intensive IT support, actually, there are many things that might be "digitized". Even if we would not describe some basic needs like ticket sale and payment processing systems, which apparently should be done using state-of-the-art solutions, we would still have some processes that can and should be automated. First, while touring the circus itself is an autonomous village; therefore, it needs to ensure that all the communications work fine. These range from the logistics (as they usually have to transfer over 50 trucks with equipment to distant locations) to the technical documentation to assemble this at some remote place. Moreover, human resources are the biggest concern for the circus. As the recruiters hire people all around the globe, they have to store huge amounts of data about potential artists including videos of their performance, over 50 costume measurements, etc. Before the transformation, all the data about the costumes (and the circus possessed several thousand of them) was stored in separate Excel files processed by many unrelated applications. Even though the personnel tried to keep the information organized, it was typically stored in differently coloured folders, which, concerning the amount of data, was very inefficient. By 2000, the IT infrastructure of the Cirque became totally chaotic. IT managers had to spend several days before each show to configure the heterogeneous software platforms for the servers. These included both Microsoft Windows and Novell Netware, a few network standards, and over 800 proprietary applications and software packages. Moreover, most of those software products were standalone, as various shows were independent businesses rather than parts of a single organization. In other words, the absence of standards and interoperability resulted in huge downtime, and excessive installation and maintenance costs multiplied by the system inefficiency. In addition, software users wasted much time before every show to get familiar with the new applications [15].

Crisis management plan

Danielle Savoie, the CIO at Cirque du Soleil, after 3 months of the IT infrastructure assessment, developed the following plan [4]. First, the new system, considering the nature of the Cirque, should use a few open interdependent and interoperable technologies. Of course, a single "platform" technology, even such a powerful one as SAP, would not support all the needs. However, the main business processes were integrated by the SAP at that point; therefore, it was decided to use SAP as the main integration channel. Danielle developed these three main objectives:

(1) Develop a vision.
(2) Find a professional IT group in order to build a system.
(3) Convince the Cirque executives that the IT transformation is mission-critical.

Discussion

Most likely, the key challenge of the Cirque, as of 2000 when this case study happened, was the lack of standards. Even with single-supplier software (e.g., Microsoft Word 2007 and 2018), and backward compatibility support, maintenance issues are quite possible. While not mission-critical for certain application types (such as text processing), some incompatibilities, in the others (for instance, accounting software) might lead to such crisis triggers as system failure, data loss, or even business downtime. As the "corona crisis" demonstrated by the recent show business revenue dropdown, the entertainment industry should be digitally transformed.

Circus business is quite challenging in this respect as watching the performance on a home TV screen is a completely different experience than a live show; one would probably refrain from paying for this "online version". However, we suppose that smart and innovative technologies might solve this issue. For instance, extended reality (XR) and particularly virtual reality (VR), would give a better immersive effect than watching the show on a flat screen. Even though good VR headsets are rather expensive, they become more and more affordable. Therefore, we would suggest extending these show features, e.g., multiple 360-degree cameras.

Moreover, for non-IT-intensive businesses, it seems reasonable using cloud services (such as AWS by Amazon or Microsoft Azure), as in many cases, this is a cheaper and more robust option than purchasing and maintaining hardware. Besides, cloud services scale better, so businesses can easily pay for more performance when required and get it instantly.

3.3 Keeping Data on the Go: Dropbox

Background

Dropbox is a software company founded in 2007 by Drew Houston and Arash Ferdowsi. It provides cloud storage and file synchronization services [28]. Users can synchronize the Dropbox folder across their devices and view their files in the web version of Dropbox. Dropbox supports all major platforms including Windows, macOS, Linux, Android, and iOS. Currently, more than 600 million people use Dropbox, and more than 500 thousand teams rely on Dropbox Business.

The Idea of DropBox as the seamlessly synchronized cloud file storage belongs to Drew Houston. Once, when he was travelling, he realized that he forgot his USB stick with the data he needed for his work. Therewith, the idea of a service which provided the synchronized shared access to files between different computers came to his mind. After he felt the need for such service, he quit his job and recruited Arash Ferdowsi, who dropped out of MIT. Arash later became Dropbox's co-founder and chief technology officer.

For the user, DropBox works like a shared folder between multiple devices. After users put files in the DropBox folder on their device, it comes available on other devices. The key feature of DropBox is a unique binary differentiation algorithm allowing seamless file synchronization between multiple devices. In an IT engineer's view, this digital product is a toolkit similar to git and trac with an advanced user interface.

Walkthrough

DropBox as a company developed in an original way. In April 2007, Drew and Arash started the company and by September 2008 the first beta release was ready and launched. Dropbox received its first financing through the Y Combinator program. After that, it was supported by the Sequoia Capital venture firm.

At the early stages of the development, Dropbox was available to a few users and its founders reached investors through online videos demonstrating the product. The same technique worked to get over 75,000 registrations for their beta testing program.

To attract people for the beta test launch, Houston took an unusual decision. He recorded a 3-min video about DropBox and posted it on social media for geeks. That boosted the number of beta testers from 5,000 to over 75,000 [10].

Drew and Arash decided to use freemium for the business model. That is, when certain functions are free while the advanced features are paid [11]. For the beta testing, the payments were not available. Fortunately, the beta testing gave a large amount of useful feedback about their project. Houston and Ferdowsi fixed all the critical issues found in beta testing and went to Y Combinator. There they had successfully raised money for project development and growth [10].

The project progressed, and the number of users grew. The only problem was the low number of paid users. The DropBox team tried to solve this problem by hiding the free solution for those who came from AdWord; however, this did not work. Moreover, the company had to spend USD 300 for each paid user, while the annual payment was USD 99 only. Consequently, they started investigating the problem and hired analysts who found out that DropBox's audience grew spontaneously.

After that, the team decided to make a referral program and rely on the "word of mouth". As a result, the 4 million users available attracted 2.8 million new users [8, 13]. In addition, the team carried out an extensive UX research. The initial results were unsatisfactory; however, they fixed over 70 critical bugs, which allowed to better manage their active users. Besides the UX, DropBox performed marketing research and found out that many free users used a "hack" of Trash folder, where the files were stored forever without any memory limits. This issue, together with other critical vulnerabilities, was fixed.

Crisis of transformation

The most likely reason for the DropBox crisis transformation was lack of paid users. However, that was a surface problem, which clearly caused the lack of profit. After a deeper investigation, the root cause problem found was their policy to provide their solution for free. In case of an inadequate solution for the problem, DropBox could easily run out of the budget as a result of this crisis. However, they managed to produce a reasonable answer, as we discuss further.

Crisis management plan

The first attempt to manage the crisis was a simple brute force solution. The company tried pushing their clients to buy their product. Naturally, that brute force approach did not work well. As the company applied the PDCA/DMAIC approach to analyse this initial failure and realized that it did not work, they started adjusting and improving. They performed a series of research, which eventually revealed the project's bottleneck. This "think over" method being a PDCA/DMAIC-based approach was an adequate solution for the crisis they faced.

Their research areas included:

- User experience (UX),
- User interface (UI),
- Marketing,
- Required features.

The old (before crisis) and new (after crisis) company state

Before the crisis, the company had a reasonable organic user growth; however, this did not bring the money in. After this issue was solved, the company focused on user growth addressing the human factors hampered by mediocre UX and UI, and thereby fixed crucial business problems. Currently, DropBox grows steadily; they keep the active users and make more profits.

Dropbox succeeded in creating product and marketing strategies that helped the company to grow. Since Dropbox chose an innovative freemium model, the team required understanding the benefit of analytics and organic user acquisition [19]. These techniques are quite common today; however, in the pre-digitalization era, they were novel. The reason why the paid advertisements did not work well, according to Drew Houston, was that "It is not like the average user wakes up in the morning wishing to get rid of their USB drive. If you do not think you have a problem, you are not going to look for a solution. Search is great for harvesting demand, not for creating it." As we mentioned before, the distribution deals were also low efficient because the Dropbox was not well recognizable at that time. However, the case study reads that after the Dropbox client base become larger, they created an open API. This strategy allowed the third-party applications to load and save the files using the Dropbox as a common and easy-to-use interface, and the company immediately received multiple requests from leading smartphone manufacturers to preinstall the Dropbox application on their devices.

Which techniques and practices worked in the crisis?

The two approaches that clearly worked for DropBox were (i) following adaptive PDCA/DMAIC lifecycle and (ii) harnessing human factors as a result of UX/UI research. These practices assisted in identifying the product/process bottlenecks and improving the cash flow in terms of bringing money in.

What did not work to manage the crisis? Why?

The alternative for the agile PDCA/DMAIC lifecycle is the rigid and straightforward brute force approach. This straightway was first tried to solve the crisis-related problem. However, it did not work well as the initial problem required deeper analysis to identify its root cause and plan, act, and adjust accordingly.

Possible suggestions for the crisis management.

Dropbox, if they did not do everything quite right, made no severe mistakes. Their approach included analysing, adjusting, and experimenting to find adequate product development and marketing strategies. The only immutable ingredient of the initial Dropbox strategy was product usability. The case study narrates that the Dropbox decided to keep its common user interface for its consumers and businesses rather than create a separate version for its business clients.

Recently, the company moved from the Amazon S3 service to its own cloud storage and currently offers Professional, Standard, and Advanced business plans with variable storage and administration features available. Since Dropbox is intensively used by a large number of users, this company will likely continue operating in future. Therewith, Dropbox should monitor competitive products (such as Google Drive, Microsoft OneDrive, etc.) using the same PDCA/DMAIC lifecycle. Actually, the company keeps track of the competitors and adjusts accordingly by continuously adding features for collaborative activities such as editing and signing the documents stored.

From the software engineering perspective, we would also recommend actively using quantitative metrics and statistics to further improve this PDCA/DMAIC approach.

Would this company survive in digitalization (i.e., change to Industry 4.0)? Why?

In our opinion, most likely, the company will survive, as the state of the technologies Dropbox currently possesses for the data processing could become mission-critical for the industry 4.0. Moreover, Dropbox could create a solution for a uniform data collection from a large number of flows including IoT sensors.

Suggest a strategy for this company's survival in the current market

For this company, digitalization seems manageable. The potential threat, however, may be the rapidly growing number of users of the services similar to Google Drive and Google Docs. Therefore, the way to survive would be cooperating with a leading mobile device producer, such as Apple or Huawei, to create a solution competitive against Google online services.

3.4 Going Eco and Embracing the World: Zara Fashion

Zara: The Story of transformation

Zara is an international fashion company, one of the largest in the world. It belongs to Inditex, the world's largest fashion group. Zara is a clothing retailer based in Spain [18]. Founded in 1975 by Amancio Ortega and Rosalia Mera as a family business, the company now is present in 88 countries and owns over 2,200 stores all over the world. The investment to open one's own boutique under this well-known brand starts from 2,500,000 rubles, which is around USD 33,550. Zara has become a pioneer in "fast" fashion thanks to its highly efficient supply chain [14, 27]. The company's ability to rapidly introduce clothes of the latest trends helped Zara to receive worldwide recognition and beat other competitors. Zara has a unique customer-centred business model that includes design and production, as well as distribution and sales through an extensive retail network.

Initially, the company's activities were focused on selling copies of the clothing of world fashion houses but at lower prices. This caused a great demand among customers and the opening of new stores in the country and then abroad did not take long to wait.

In those years, the time-lapse from the initial clothes design concept to their distribution was very long and lasted for several months. This required distributors to anticipate fashion trends, which created unnecessary difficulties and the risk of remaining with unsold goods. Ortega did not want to put up with this situation and dreamed of creating a so-called "fast" fashion, which could allow the company to quickly respond to the changing tastes of its target audience [23]. And he succeeded.

His fateful meeting with Jose Maria Castellano, an IT expert, resulted in a revolution in the global clothing industry. Castellano played a key role in the development of the company: in the 1980s, he helped Ortega to draw up a special work plan. Jose explained to Amancio that the profit was reduced due to large batches, and in order to increase revenue they needed to produce clothes in smaller quantities but more often. They hired more designers, modernized production, and changed the methods of delivery and accounting of goods, which allowed to reduce the time from the beginning of the development of the model to the beginning of its sales (now only up to 2 weeks) and instantly respond to changing tastes of customers. This plan was a huge success: things in the stores were sold out, and on the day of deliveries, which fans call "Day Z", customers literally guarded new collections. The company sticks to this policy today which allows it to remain one of the most popular stores on the market. The developed system was so effective that even Harvard researchers investigated it.

The first store outside Spain was opened in the late 1980s in the neighbouring Portugal, in the city of Porto. A year later, boutiques with Zara sign appeared in the United States, and then in France, Mexico, Japan, and other countries around the world.

All Zara stores specialize in fashion clothing and accessories for women and men, as well as children at affordable prices. Unlike many other fashion companies, they do not move production to China. Half of the products are produced in Spain, the rest comes from the countries in Europe, Asia, and Africa.

Since 1985, Zara brand has become a part of Inditex Group corporation, which also owns such well-known stores as Massimo Dutti, Pull and Bear, Oysho, and Bershka. In just 1 year, this company created about 12,000 new design models and produced over 1 billion pieces of clothing.

Perhaps someone may think that, in the stores of this brand, there are often the same type of things from season to season. There is a reason for this. The company has a department that collects customer reviews and if an item gets a sufficient number of positive ratings it remains in the store's collection unchanged or slightly modified. However, low-sold items are completely removed from production.

Oddly enough, the company spends no more than 0.3% of the annual budget on promoting Zara brand, while competitors spend up to 4%. The management believes that placement in prestigious and expensive places in the city helps to form an image and attract customers better than advertising.

In 2011, Ortega left the post of chairman of Inditex Group; he still owns a larger percentage of shares and is one of the richest people on the planet.

Zara has significantly reduced the time spent on designing, manufacturing, and distributing clothes with the help of IT. The key factor of their success lies in giving customers what they want before they even ask for it. To do that, designers in Zara constantly monitor customer needs in all the stores. Logistics and production in Zara also play the main role in the brand's growth. Zara has a high turnover of goods and short production season – goods change every 2 weeks. It helps the company to decrease its costs on warehousing for each store and reduce its business risks.

Zara uses IT extensively. Efficient communication, data transfer, and processing allow the team not only updating collections in 2 weeks (for H&M, it takes 3– 5 months to do this) but also selling "exactly what people want to buy at a particular moment".

Interestingly, analysts and design team share the same workspace. Moreover, Inditex has no chief designer position. Instead, their collections are entirely based on data analysis and teamwork.

Today, Zara faces digitalization issues as it needs to "go green". Recently, Zara has promised to sell only eco-friendly clothing that meets sustainable development principles starting in 2025.

During the COVID-19 pandemic, Zara's online sales have increased dramatically and the company has even reduced inventories and operating expenses. One of the reasons to achieve this is intensive IoT application. In particular, the brand began using radio frequency identification technology (RFID) a few years ago. RFID tags are installed on each product unit at the clothing factory. Information about the product is available at any time until it is purchased. After the RFID tag is separated from the clothing, it is sent to the recycling centre, where the product ID is erased from the system memory. The RFID tag can then be reused.

The introduction of RFID helped to simplify customer service, improve accounting for the receipt and release of goods from the warehouse, and inventory on the trading floor increasing accuracy up to 99%. Thanks to RFID, employees can quickly locate and replace certain clothing items. The radio frequency identification helps to keep track of the quantity of goods and consolidate the data about the size and colour performance of each model. RFID application halved the collection update time.

During tough economic times, some customers could find even Zara items overpriced. Therefore, they applied data analysis to optimize prices. Until 2007, cost reduction decisions in discount campaigns were manual. In 2010, Zara introduced a computer optimization model to justify the price cut down. The implementation in Belgian and Irish stores allowed increasing sale revenues by 6% after just one season. According to project experts, Zara still uses IT-intensive approaches to make decisions on the markdown of goods during sales.

The company struggles to survive in the current digitalization and will likely invest more in its online store and technology development (including the currently widespread big data) in order to become more environmentally friendly and develop sustainably. We recommend that Zara should use IT not only to reduce the time to market and understand the customer's needs but also to attract the customer's attention online and interact with them more actively. For instance, they can use augmented reality (AR) technologies for the customers to try the clothes on or even design their own items. Social media-based campaigns would also be beneficial, including live sessions and online competitions. This would help managing customer relationships, and consequently raising sales. Also, investing in smart technologies and digital channels would reduce logistics and inventory overheads significantly and assist in sustainable development.

As a vertically integrated retail chain supplier, Zara integrated design, production, and trade processes. Unlike many clothing distributors, Zara controls almost 100% of this value chain. [26].

Digital production technologies, following the customers, creating unique products, and bringing them to the market faster than the competitors give an impressive impetus to the business development. As a result, Zara is at the top of many respected ratings.

We recommend that the company focuses on AR technologies to survive in the current digital transformation. These technologies assist in designing online realistic product mockups and therefore reduce the percentage of returns to the retailers. For example, a similar iOS application of the Russian Lamoda fashion company features a virtual AR fitting room. In general, AR creates new digital experiences that enrich the relationships between the consumer and the brand thus harnessing human factors to provide sustainable development.

After analysing the Zara company transformation, we can draw the following conclusions:

(1) Company produces clothes entirely by itself, starting from concept design. Zara produces around 50% of its collections at the beginning of each season, and the rest during the season.
(2) Company has their own production facilities.
(3) Company enjoys a relatively stable and developing market.

3.5 Locating You Anywhere: Foursquare

Foursquare is a location technology platform offering business solutions and consumer products through a deep understanding of location. It began as a check-in application in the late 2000s. It was originally founded by Dennis Crowley and Naveen Selvadurai in 2008 and launched in 2009 [9]. This happened after Crowley's initial location-based Dodgeball startup was acquired by Google in 2005 and shut down 4 years later.

With 60,000 users and 10 employees in the fall of 2009, the company now has 55 million users and about 400 employees [22]. To date, the company has raised an amount of over USD 390 M in funding [1, 20].

History and early development

The company's genesis stems from a thesis written together by Dennis Crowley and Alex Rainert while at New York University (NYU) in the early 2000s. The thesis described Dodgeball, a platform that allowed users to share their whereabouts by sending a venue's address in a text message to Dodgeball's servers. Dodgeball then matched the address to a location in its directory and texted the users' friends and friends of friends within a ten-block radius[21]. Users could also "shout" messages to all of their Dodgeball friends. Dodgeball was developed as an application after Dennis and Alex graduated from NYU, and they sold it to Google afterwards. Unfortunately, Google shut down dodgeball in January 2009, 4 years after its acquisition.

When the iPhone came to light in 2007, Apple introduced the iPhone software development kit (SDK), that allowed developers to build applications that utilized the phone's hardware and software. Dennis Crowley teamed up with his friend and computer scientist, Naveen Selvadurai, to build a location-based service that had more functionality than Dodgeball.

The alpha version of Foursquare was an iPhone application and a website that was available to 80 of Dennis and Naveen's friends in New York, San Francisco, and Los Angeles [16]. The main features at the time were to check in to any public establishment and view nearby locations. They added more social and game-like features in the subsequent releases, such as receiving points for checking in to new places, tips about locations, and seeing other people's check-ins [3].

Transformation and crisis

The launch of Foursquare coincided with the SXSW, a large annual confab for the online community held in Austin, Texas. Foursquare gained thousands of users during this period, even though the early releases of the application were available for iPhones only. It can be said that the SXSW conference in 2009 greatly increased the popularity of Foursquare, as it got popular websites such as Mashable talking about it among various other media buzzes [25].

However, the media buzz stopped soon after this event, and Foursquare's growth stalled during the summer of 2009.

It was time to seek funding as up till that point, funding had been by Crowley and Selvadurai. It was challenging to pitch the service to investors. This was because Foursquare's business model was ambiguous. Additionally, it was difficult to differentiate Foursquare from competing startups with similar features.

Managing the crisis

Foursquare did a few things to help manage the crisis it went through and to manage the stalled growth it experienced.

The first solution Foursquare adopted was expanding its mobile platform. Foursquare capitalized on the growing popularity of smartphones at the time and extended its application to Google's Android operating system, and soon after, released a Blackberry version of the app. The company also released a new version of its iPhone app that allowed users to see profiles of other users who had checked in to the same venue but were strangers [2]. Also, they applied social media platforms such as Twitter to get the customer feedback on the new apps, and this helped improving [5]. Crowley attributed Foursquare's growth to the proliferation of smartphones and broad acceptance of multiple social networking platforms.

Additionally, Foursquare rewrote its API from the initially internal version and made it public. By the fall of 2010, over 1,500 developers used Foursquare's API to develop third-party apps, and this also helped the company.

State of company after crisis management—The organization

The growth of Foursquare sprung up again after they were able to manage the crisis. With the rapid user base growth, Crowley and Selvadurai needed to adjust the company management style [2].

The first step to that was the hiring of highly skilled engineers. They hired a former Google employee as their lead engineer, website developer, and server-side architect. They also hired several mobile app developers on contract basis [1].

Other business development hires followed: they brought into the team a general manager, a marketing director, a lawyer, and a business development team. An organization has indeed been built. Crowley became the CEO of the company.

The team grew, and by the fall of 2010, Foursquare had 45 employees, of which more than half were engineers. The company had a five-person product group that included a product manager along with user experience and design experts.

State of company after crisis—The product

After the challenges and crises, the application was significantly upgraded and improved. The app was expanded to 100 cities worldwide, and new features were introduced. These included specific badges for each city they targeted, which were publicly available on the user's profile [17].

Another add-on was promoting venues by the Specials feature so that the app could provide tips, suggestions, and notifications on Special deals in the neighbouring locations.

In January 2010, the company launched "foursquare Everywhere" based on the user's feedback. This allowed users to sign up for the service from any city in the world. Also, Foursquare introduced the new app features that allowed individual users adding new locations and businesses claiming their locations.

Foursquare API became widely used and applications such as games, email services, and other location-based apps were built on top of this API.

Foursquare and Industry 4.0

Foursquare as a location technology company working with a deep understanding of user location, is well positioned towards the digitalization and the transformation to industry 4.0. In the beginning, when the idea was just about check-ins and "shouts" to friends about locations, the chance for the company to survive digitalization was low. However, the situation changed dramatically with the advent of third-party SDKs and the mobile device proliferation, which Foursquare took advantage of.

On-demand availability of computer system resources, smart sensors, mobile devices, and location discovery technologies, are all a part of the digitalization technology stack. Solutions such as the Foursquare City Guides, the API with advanced user location of the metadata information among others, make it very well fit to be able to stand the era of Industry 4.0.

Foursquare has gone through very extensive phases of transformation from the ideas of Dodgeball which relied on SMS, to the development of city guides, tips, reviews, the concept of turning life into a game with real-time points and rewards for user movement, to the development of APIs and advanced mobile and web applications that provide a stepping stone for others to build intelligent applications deeply rooted in location data.

The team has been able to survive over a decade of transformation and is now one of the top providers of advanced user location platforms in the world. After acquisition of Placed from Snap Inc., Foursquare is further expanding to gain a better position in the digitalization contest.

3.6 Mass Entertaining Ubiquity: Disney

Disney: The history of transformation

The Walt Disney Company is one of the largest entertainment media conglomerates in the world today. The studio was founded on October 16, 1923, by brothers Walter Elias and Roy Oliver Disney. At first, it sold its handwritten drawings (so-called stop-motion pictures), then it was one of the first to produce colour animated pictures. Disney cartoons became popular and were among the first to win Oscars. The studio is currently one of the most popular content producers and media conglomerates. As part of its subsidiaries, Disney has gathered under itself many well-known and large-scale studios: Walt Disney Animation Studios, Pixar, Marvel Studios, Lucasfilm, 20th Century Studios, Blue Sky Studios, and a majority part of 73% in National Geographic. Also, in addition to the studios, Disney has more than 11 world amusement parks (the most famous is Disneyland), radio networks, television broadcasting, resorts, merchandise stores, and so on. In 2019, the company's capital was USD 311 billion, and the net profit was about USD 11 billion. The profits and stock prices grew, but in recent years, there has been a dramatic change in the way of life that could not but affect Disney. In 2020, the Coronavirus crisis occurred.

It would seem that Disney is a company that depends very sensitively on the live presence of people. Premieres are shown in movie theatres, people want to visit the theme centre in person, and actors are afraid of getting sick on the set. Virtually, everything the company does must freeze because of the pandemic. However, the company's management and employees have been able to cope with these problems by digitalizing their activities. These changes are discussed further in this section.

Cinema's prohibition. What the film industry should do?

In 2020, due to COVID-19 restrictions, all cinemas were closed for a long time. All premieres lost their box office receipts, which are the main source of income for continuous movie production. Disney owned about 40% of the USA box office, largely due to the success of the Marvel movies. However, the pandemic halted revenue growth in most of Disney's business areas.

Seven months after the pandemic began, the company decided to rethink its business model and bet on developing a streaming service. They promptly designed and developed Disney Plus, a digital streaming service to bring content home to the customer. The idea of delivering content, instead of the conventional cinema shows, looks ordinary to the user. However, for the company producing the content, this required changing their operating and business model. Walt Disney announced plans for a wide restructuring of its entertainment and media departments, with the aim of developing its D2C strategy (i.e., production, promotion, and sale of content directly to the consumer, without intermediaries) and its streaming division.

Before the announcement of the business model change, most stock market analysts predicted a halt to Disney's business. They believed that the company was waiting for the end of the pandemic to return to its previous activities.

However, the new business model brought a significant market advantage and put Disney's competitors in a vulnerable position. The company's willingness to deliver content as a streaming service rather than holding on to the old-fashioned format allowed Disney to attract customers and keep profits when all other movie theatres were closed.

The growth rate of Disney Plus could affect the performance of traditional cable TV companies. The increase in subscriptions to the new service could decline cable TV popularity. Moreover, Disney would increase their customer base due to the following factors:

(1) Content is easier accessible for home TV.
(2) With borders closed, customers spend more time at home looking for new entertainment.
(3) Subscription-based access is more convenient as the clients get a rich set of products for a fixed price.

A year after the launch of the Disney Plus service, The Walt Disney Company management decided to restructure, focusing on streaming and direct content sales. This included forming a department for distribution of media and entertainment content. Obviously, this restructuring triggered the company's growth by 40–50% in 2020, while the cinemas were closed in the COVID crisis.

In the short term, the reorganization effect is to multiply, Disney's financial statements reflecting the update as early as the first quarter of fiscal 2021.

Applying Christensen's business model (see Fig. 3.1) to Disney, let us examine the four key business model components.

Value Proposition, i.e., media content. In the media network segment, digitalization revolutionized distributing and consuming content among the clients of Disney's television and cable networks that are part of Disney. The largest global target audience for Disney Channels' television channels is the Millennial generation, i.e., people born in 1987–1997. Millennials differ from previous generations because from an early age of their childhood, they have been exposed to digital technologies. Consequently, they are used to consuming content in massive amounts, at any daytime, anywhere, and, most importantly, through any device. Nielsen Research states that millennials aged 18–24 make up to 98% of most active smartphone users, and their content consumption is often up to 18 hours per day. These figures prove that the business model of media companies should adequately address their customer values and expectations concerning digital content. Disney's business solution is precisely aimed at this focus group as its content is easily accessible across multiple kinds of devices, and nearly infinite in volume [24].

The Profit Formula. One of Disney's key problems related to the Industry 4.0 technological shift, is the widespread of the Internet, and consequently, piracy. The

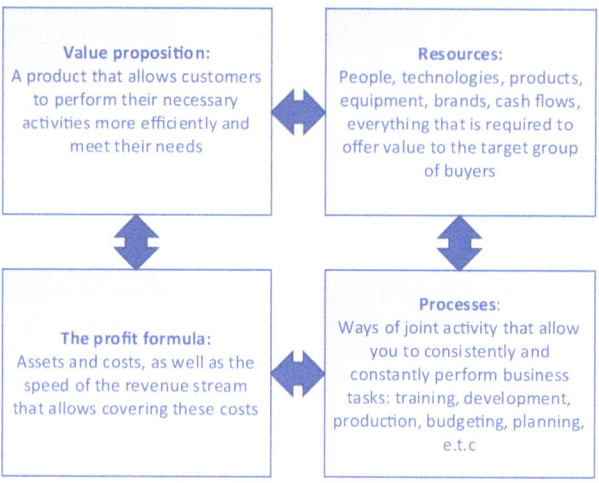

Fig. 3.1 The business model (by Christensen, Johnson, and Kagermann)

increase in the mobile data transmission speed and the digital technology development allowed copying and distributing the company's intellectual property assets such as movies, online videos, TV shows, etc. Disney annually spends a tremendous amount of money on lobbying for copyright law in the USA: since 1997, the total investments amount to USD 87.6 billion. Digitalization has dramatically increased this value.

Resources. The ubiquity of digital technologies, especially mobile platforms and social media allows ordinary users to reinforce or destroy brands and company reputations, which is an important success factor of Disney's business model. User-generated content allows bloggers, freelance journalists, or social activists to express their opinions about Disney products or the company's market and public behavior. Disney implements various communication strategies. They actively use storytelling through different mobile platforms to draw customers' attention to the brand. Most of the related Twitter posts can be divided into two groups: tweets aimed at promoting Disney products and those related to the current social trends. At the same time, Instagram shows a more "human" side of the company: the posts are related to the everyday life of Disney employees, charitable activities, environmental issues, etc.

Processes. The negative impact of digitalization is reflected in the company's cybersecurity risk management processes. In particular, according to the annual report, Disney stores all data digitally, and any data theft or corruption that could have a negative impact on the company's reputation, adds additional costs for recovery or result in a significant loss of profits, especially in the case of commercial and/or confidential information. The positive impact of digitalization, for example, is that a company can use the power of artificial intelligence to automatically distribute content to their multiple platforms, which can substantially reduce the cost of maintaining business processes.

As an analysis result of Disney's digitalization, we can conclude that the company has successfully applied the transition to the new smart business model and continues to move in this direction. The change in the model allowed them to survive in difficult times, and in our opinion, the company will not revert to the old operating mode. Harnessing big data and computing power promoted a large-scale streaming service. At the same time, the business model change allowed the company to attract a larger number of new consumers, retain loyal customers, and increase profits.

3.7 Conclusion

This chapter analysed a set of earlier, i.e., pre-digitalization transformational cases for large-scale businesses, many of which were also IT-intensive. The key hampering factor detected was multidimensional complexity of their operational landscape; this included technology, business, and human-related factors. As this complexity might critically affect their digital transformation in a dynamic and fluctuating environment, and trigger a crisis, we used the ITC framework. Careful application of the

ITC-based case method highlighted a set of critically important aspects such as business diversity, organizational structure, and geographical variety, to name a few. The case studies embraced different professional areas such as consulting, show business, IT development, and fashion industry. As a result of ITC-based case method application, the strong and weak sides of the digital transformations were identified, and sustainable development strategies and techniques were outlined. Thereby, the approach succeeded as a concept proof and provided directions for further research.

References

1. Ali, R. (September 4, 2009). Social App Foursquare Takes in $1.35 Million in Funding from Union Square, paidContent.org., Retrieved from http://paidcontent.org/article/419-socisl-app-foursquare-takes-in-1.35-million-in-funding-from-unionsquare (Last Accessed February 17, 2022).
2. Bruno, A. (April 24, 2010). Q&A: Foursquare CEO Dennis Crowley, Billboard. via Factiva.
3. Can Foursquare and Gowalla Put Mobile Ads on the Map? Neto Media Ase, June 10, 2010, via Factiva, accessed September 2010.
4. Cirque du Soleil, A Fantastic Journey. Retrieved from http://www.cirquedusoleil.com/cirque dusoleil/pdf/pressroom/en/historique en.pdf. (Last Accessed February 17, 2022).
5. Cirque du Soleil, Cirque du Soleil at a Glance. Retrieved from http://www.cirquedusoleil.com/cirquedusoleil/pdf/pressroom/en/cds en bref en.pdf, Last Accessed February 17, 2022).
6. Dragoon, A. (May 21, 2005). The Amazing Traveling IT Show, CIO, Vol. 16, No. 3, November 1, 2002.
7. Dybwad, B. (Jan. 14, 2010). Twitter Growing Internationally, Mashable. http://mashable.com/2010/01/14/twitter-growing-internationally. (Last Accessed February 17, 2022).
8. Gannes, L. (Nov. 24, 2009). Dropbox raises $7.25M, Crosses 3M Users, GigaOm blog, Retrieved from http://gigaom.com/2009/11/24/dropbox-raises-7-25m-crosses-3m-users/on (Last Accessed February 17, 2022).
9. Gustin, S. (September 2, 2009). Hip to be foursquare: A Nnght out with CEO Dennis Crowley, Daily Finance. Retrieved from http://www.dailyfinance.com/story/hip-to-be-foursquare-a-night-out-with-ceo-dennis-crowley/19147931, (Last Accessed February 17, 2022).
10. Houston, D. (April 23, 2010). Customer Development Case Study: Dropbox, video file, Justin.tv, Retrieved from http://www.justin.tv/startuplessonslearned/b/262672510. (Last Accessed February 17, 2022).
11. Houston, D. (April 16, 2010). Drew Houston: Freemium for Consumer Internet Businesses, Part 1, video file, YouTube. Retrieved from http://www.youtube.com/watch?v=TBTyjBQ9Eq4. (Last Accessed February 17, 2022).
12. Houston, D. (April 16, 2010). Drew Houston: Freemium for Consumer Internet Businesses, Part 2, video file, YouTube. Retrieved from http://www.youtube.com/watch?v=9I30YUSb568 (Last Accessed February 17, 2022).
13. Houston, D. (April 16, 2010). Drew Houston: Freemium for Consumer Internet Businesses, Part 3, video file, YouTube. Retrieved from http://www.youtube.com/watch?v=WOOYqK1qE3g. (Last Accessed February 17, 2022).
14. How Zara Fashions its Supply Chain, Strategic Direction. (2005). Vol. 21, No. 10, pp. 28–31, Emerald Group Publishing Limited.
15. "IT Strategic Plan—Stage One," presented at the Executive Committee meeting by Danielle Savoie, July 11–13, 2000.
16. Lowensohn, J. (April 9, 2009). Q&A: Foursquare Co-Creator on Privacy, Easter Eggs, post on blog "Webware,". Retrieved from http://news.cnet.com/8301-17939_109-10215732-2.html, (Last Accessed February 17, 2022).

17. Loveland, M. (July 15, 2011). Foursquare Ups the Price of Custom Badge Program. Sribbal, Retrieved from http://www.scribbal.com/2011/07/foursquare-ups-the-price-of-custom-badge-program-75k-commitment-required/ (Last Accessed February 17, 2022)
18. Murphy, R. (2008). Expansion Boosts Inditex Net, Women"s Wear Daily, April 1.
19. "Meet the Team" interview on Dropbox blog. (Feb. 9, 2009). Retrieved from http://blog.dropbox.com/?p=23 (Last Accessed February 17, 2022).
20. Rob, P. (February 28, 2010). "Hip to be Fourse arsquare, a" Washington Fost, via Factiva.
21. Sohn, A. (n.d). Newtork [sic] Rivalry, New York, Retrieved from http://nymag.com/nymetro/nightlife/sex/columns/mating/11515 (Last Accessed February 17, 2022).
22. Snow, S. (May 24, 2010). Inside Foursquare: Checking in Before the Party Started (Part I), Wired. Retrieved from http://www.wired.com/epicenter/2010/05/inside-foursquare-checking-in-before-the-party-started-part-i/all/1, (Last Accessed February 17, 2022).
23. Sull, D., & Turconi, S.(2008). Fast fashion lessons. *Business Strategy Review*, Summer.
24. The Walt Disney Company Enterprise Architecture Overview. Uploaded on Sep 12, 2014, Retrieved from https://www.slideserve.com/samira/the-walt-disney-company-enterprise-architecture-overview (Last Accessed February 17, 2022).
25. Van Grove Jennifer. (March 16, 2009). "Foursquare Is the Breakout Mobile App at SXSW," post on blog "Mashable,". Retrieved from http://mashable.com/2009/03/16/foursquare. (Last Accessed February 17, 2022).
26. Walters, D. (2006). Effectiveness and efficiency: The role of demand chain management. *The International Journal of Logistics Management, 17*(1), 75–94, Emerald Group Publishing Limited.
27. Waller, A. (1998). The globalization of business: the role of supply chain management. *Management Focus,* (11).
28. "Worldwide Online Backup Services 2007–2011 Forecast: A New Market Emerges," IDC, December 2007.

Chapter 4
Fostering Digitalization: IT-Intensive Businesses

4.1 Food Industry: Dodo Pizza and Drinkit

A coffee house is mostly a traditional business with long-term stable processes. For example, in the morning the client comes to the coffee house, stays in a queue, talks with the barista, orders a coffee, and drinks the coffee inside or to go. Every time the barista has to repeat the menu, suggest promos and other services, repeat orders after the whole process, and wait for a check. On the other side, the client waits in a queue, waits for the barista talking, long thinking and choosing, and waits for a check. This sequence of processes takes a lot of time from the barista and client lives which they can spend for themselves.

Drinkit creates a dramatic innovation in the barista-client communication process by introducing a new process for an order. The company transfers the whole process to the interaction only between client and the app. Clients interact with the barista only to get coffee from hand. Drinkit digitalizes every stage of traditional process which helps them to reduce the time which cashier and barista spend talking with a client, the client waiting time, add full customization of coffee in one place (inside the mobile app), and add user-friendly menu and payment.

The questions highlighting the key points of this case study are:

1. Why do traditional businesses digitalize? What are changes in the serving order processes [2]?
2. How to increase retention through high-client customization?
3. How does Drinkit differ from its competitors? Which are the next steps in Drinkit digitalization?
4. Is process transparency important in business? Which pros and cons can you list?
5. How do you feel about Gemba practice? Why? Is it useful? How does the Drinkit management use Gemba in their business processes?
6. Does the Drinkit digitalization improve their client relationships? Why?

© The Author(s), under exclusive license to Springer Nature Singapore Pte Ltd. 2022   65
S. V. Zykov, *IT Crisisology Casebook*, Smart Innovation, Systems
and Technologies 300, https://doi.org/10.1007/978-981-19-2231-2_4

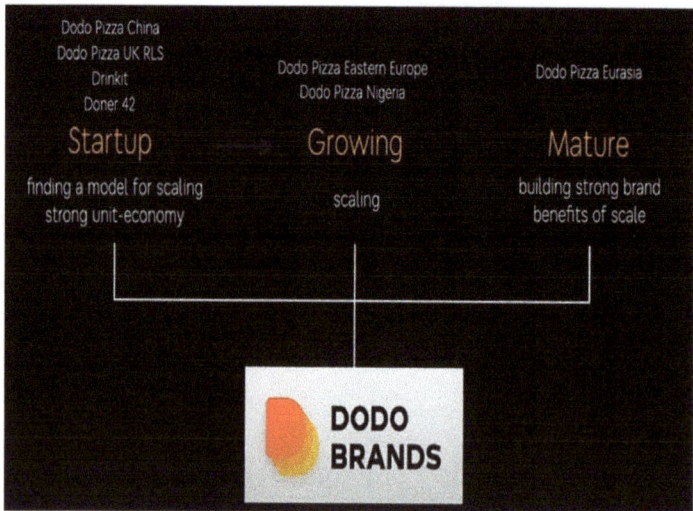

Fig. 4.1 Dodo Brands business structure

Overview of the company

Drinkit [10] is a startup company within the Dodo Brands group of companies [11]. (Fig. 4.1).

The business team started working on the project from the beginning of 2020; this included designing the concept, key ideas, economics, and menu. The application development began in March 2020, and the coffee shop itself opened in August [19]. By November 2020, the Drinkit coffee house was a room with a mini kitchen and seat places located on the ground floor of the Omega Plaza business centre in Moscow, Russia (see Fig. 4.2). Drinkit built its information system on top of the Dodo IS, i.e., Dodo information system [20–22].

Drinkit: the concept

A coffee shop, selling products offline, acquires a digital platform, typically framed as an application. The coffee shop customers are able to plan their visits, and order the desired coffee online, at the right time, with the options of a drink constructor and extras.

The application has three interacting sides:

- Client,
- Barista,
- Coffee shop manager.

Further, we describe user stories from the perspectives of the client, barista, and coffee shop manager in order to better understand the changes in typical user behaviour patterns.

Fig. 4.2 Elizaveta Schwets (IT Brand Magic in Dodo Brands [23]) in Drinkit, Omega Plaza

Concept from a customer perspective

User Stories

User Story 1

Ivan works in the office. As usual, in the morning at 8, he runs late for work. Every morning Ivan walks past the Drinkit coffee shop [13]. In the subway, Ivan takes out his smartphone and opens the Drinkit app, in which he chooses coffee for himself, makes it as large as possible, adds mint syrup and sprinkles, chooses toast and cheesecakes with sour cream. Then Ivan pays for the purchase. Approaching the coffee shop, Ivan receives a notification that his order is ready. Ivan enters the coffee shop and picks up the order without a queue. Ivan runs further to work. Ivan gets to work on time and drinks a sip of fresh coffee from his energy source. Ivan decides to write a review about the coffee shop and praise the timeliness of order processing. Ivan opens the Drinkit app, selects feedback option, and sends it. Some 15 min later, Ivan receives a response from Drinkit about his review which thanks him for visiting the coffee shop, together with a personal promotion carefully matching his previous choices, in the next order.

User Story 2

Sarah is a creative person. She goes to Drinkit every night. Sarah wants to find for herself the most non-trivial coffee tastes that she will like. Coming to Drinkit, she exchanges a few words with the barista and sits down at the table. She opens the Drinkit application, selects coffee, and then carefully picks out syrup, milk, and sprinkles, which, as she thinks, will make a very tasty combination. After making her choice, she immediately pays through the application and waits for the order.

She monitors the order progress in the application and on the coffee shop screen. After the order is ready, she picks up the coffee and drinks it. Then she talks a little with the barista about the coffee and its various tastes, about life, and goes home. On the way home, Sarah opens her shopping history in the app and understands that she has already bought 20 cappuccinos with various combinations of milk, toppings, and syrups. Sarah decides to publish an Instagram post with her favourite combinations. In the post, Sarah mentions her Drinkit account. Half an hour later, Sarah receives a comment from the Drinkit account in the Instagram, and in the Drinkit app she receives a personal promotion for cappuccino in the next order with a combination suggested by Drinkit's Instagram account.

The options of Drinkit application for the client [17]:

1. View menu,
2. Select product,
3. Customize product,
4. Share application,
5. Pay online,
6. Write feedback.

The screenshots below illustrate the app functions (see Fig. 4.3).

Concept from a barista perspective

User stories

User Story 1

Sergey is a barista. Today he goes to work again. As usual, in the morning, he sets up the coffee machine and prepares cups and lids. An order comes in for a double cappuccino with cherry topping and wild berry toast. On the tablet above the coffee machine, he receives an order for coffee. Sergey presses the coffee plate and starts cooking. At the same time, the order progress bar appears on the customer's phone and on the screen in the coffee shop. After Sergey makes coffee, he adds sprinkles and taps to confirm it is ready. At this moment, the toast barista finishes the toast and also presses ready. Coffee and toast meet together at the checkout and the cashier taps on the tablet to confirm the order is ready for pickup. The client comes for the order, and the cashier confirms that the order has been processed.

User Story 2

Ulyana is a toast maker. As usual, in the morning Ulyana makes the preparations required. An order comes in for a double cappuccino with cherry topping and raspberry toast. Ulyana presses to start cooking on the tablet above her workplace. Cooking progress starts. After Ulyana makes the toast, she wraps it up, puts it on the checkout, and clicks "Finish". This completes the toast-making process.

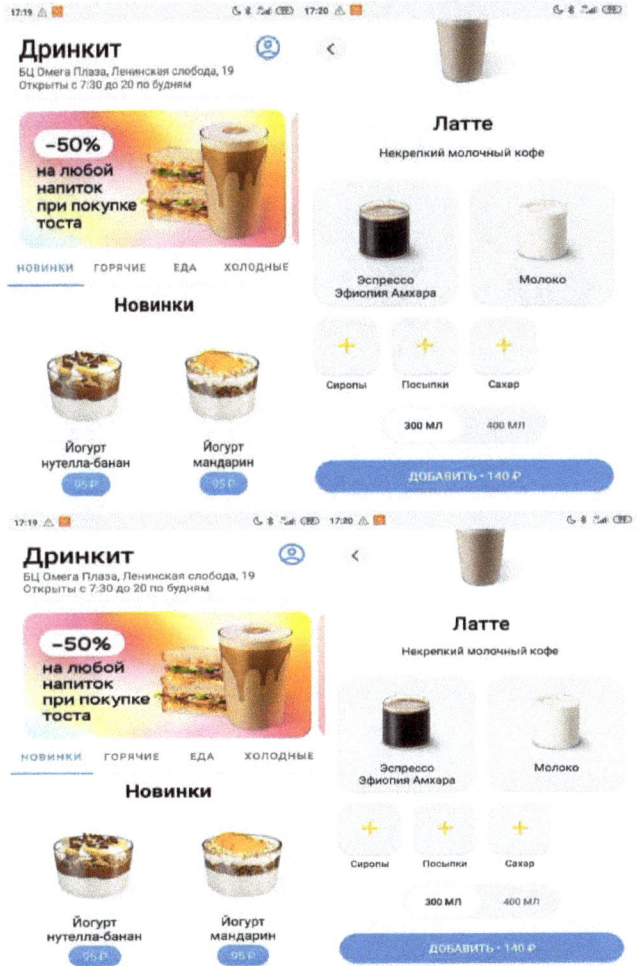

Fig. 4.3 Screenshots from the app

The ordering system for the staff features the following:

- Complete order chain.
- Personalize order chain for each area in the kitchen such as toast, coffee, and cashier.
- Track time and cooking stages.

  Figures 4.4 and 4.5 illustrate the cooking process.

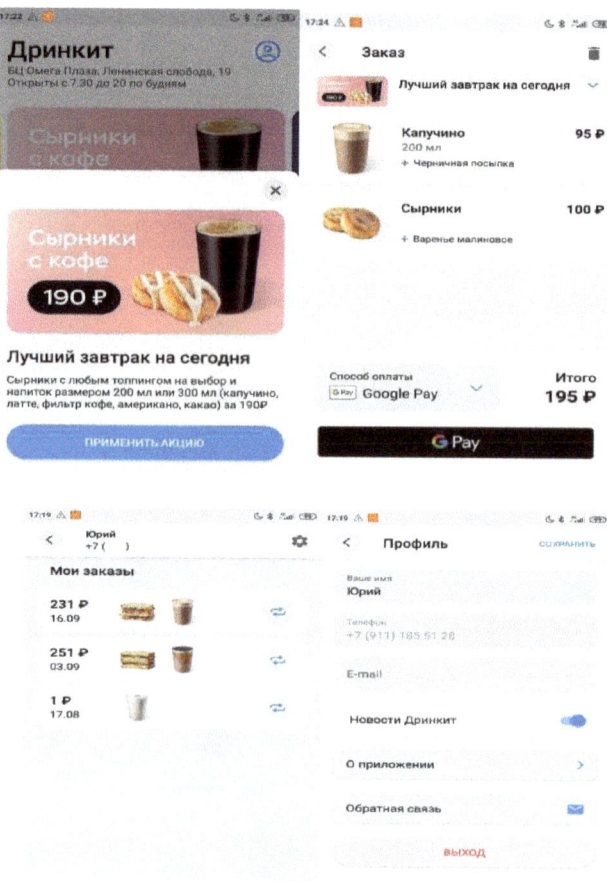

Fig. 4.3 (continued)

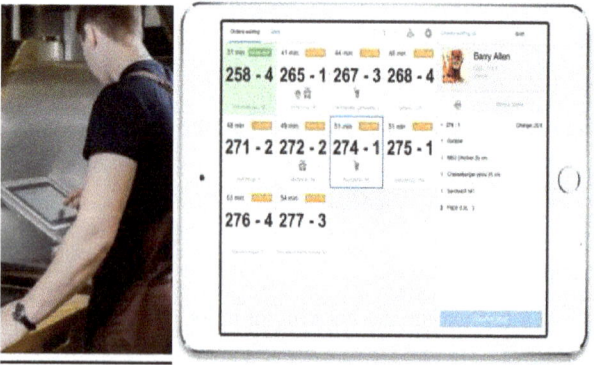

Fig. 4.4 Tablet with the orders in the cashier stage

Fig. 4.5 Tablet with the order management system

Concept from a coffee house manager perspective

User Story

Anna is the manager. After the coffee shop closes at the end of the day, Anna immediately sees the entire shop order dynamics on her tablet. These include the orders made through the application and checkout, average time to complete order, process bottlenecks, and the most intensive workers who require a day off. After a hard day, Anna discusses the processes with her colleagues to identify the achievements and the points to improve.

Figure 4.6 presents a daily analytics screenshot from the coffee house management system.

Открыта 14.08 05:40   Закрыть смену

	Кол-во	Сумма, ₽	Наличные, ₽	Карта, ₽	Онлайн, ₽
а №1	458	42 629,00	1 360,00	6 990,00	—
	0	0,00	0,00	0,00	—

Fig. 4.6 Daily analytics

Drinkit team

From the very beginning of the startup, it was laid down that every employee discusses any issue extensively, each individual is responsible for the life of the coffee shop, and everyone has the potential for growth. Anastasia Nikitina and Anna Repo (founder and manager) personally work in different positions. They do this in order to work out better all the processes in the coffee shop, improve practices, and monitor the environment from the fields. They follow the Gemba training, i.e., top managers practicing as kitchen workers and other employees to detect process bottlenecks and adjust the business procedures. For example, in the process of work, Anastasia detects the inaccuracies of the application and notices the fact that some guests pick up orders late after they have been prepared, so the orders cool down. Anastasia understands that this problem needs to be solved (Fig. 4.7).

The team undergoes continuous training; they spend time together after work to discuss the achievements and issues.

From the project kick-off, the entire business process was open, and the founder posted each step on Instagram. This created a unique atmosphere of openness around the project which is generally uncommon for the Russians. Another rare positive example of creating such a warm business climate was the blog by Fyodor Ovchinnikov, the Dodo Pizza CEO [6, 14]. Every day Anastasia Nikitina publishes a report on the Telegram channel [17, 18]. The company praises quality in all its manifestations, including delivery time, best ingredients, and shareholder relationships (Figs. 4.8 and 4.9).

Drinkit process

Currently, there are over 35 chains of Russian coffee shops and cafes with officially registered apps.

In their comments, many customers state that after establishing the application, the coffee shop moves away from the client as the technologies separate them. However, the founders often reply that digitalization revolutionizes the culture

Fig. 4.7 Anastasia Nikitina (right) working together with the coffee house staff [9]

Fig. 4.8 Anastasia Nikitina
studying with her team

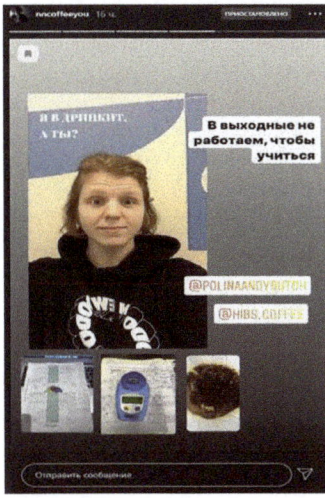

Fig. 4.9 Anastatiya Nikitina
and her Instagram posts on
Drinkit [13].

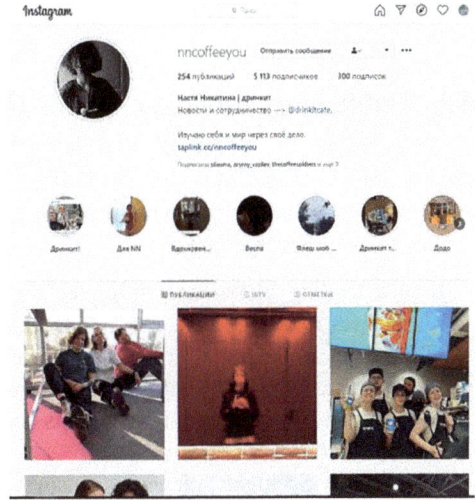

through improved services, and consequently this change is only for the better. After all, accepting orders from the application adds no critical constraints. Conversely, the shops remain open to clients and happy to communicate, and their employees remain interactive. A digital coffee shop's aim is receiving 100% of orders through the application. However, baristas still need to inform customers on the varieties of the drinks and desserts in a clear and attractive way. Therefore, the bar and dispensing area should meet these interactive requirements (Fig. 4.10).

Fig. 4.10 An Instagram post illustrating how Drinkit cares for its clients [8]

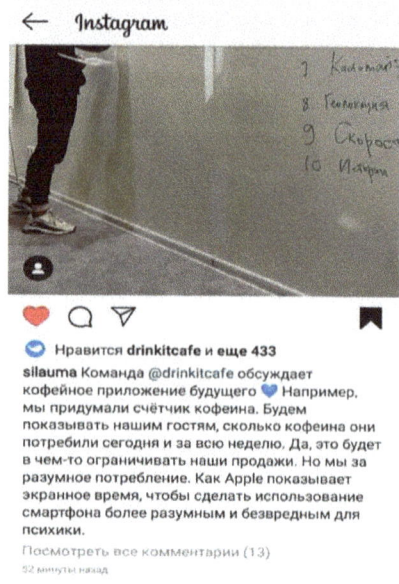

4.2 Publishing: Springer Versus IGI Global

On his 25th birthday, May 10, 1842, Julius Springer founded his bookstore and publishing house in Berlin, laying the foundation for today's company.

Key facts of Springer are:

- It is a leading global scientific, technical, and medical publisher, providing researchers in academia, scientific institutions, and corporate R&D departments with quality content via innovative information products and services.
- Trusted local-language publisher in Europe – especially in Germany and the Netherlands – primarily for physicians and professionals working in health care and education.
- Turnover of Springer Science + Business Media in 2013: approximately EUR 943 million; in 2012: approximately EUR 981 million; in 2011: approximately EUR 875 million.
- Around 2,200 English-language journals and over 8,400 new book titles published in 2013, in five main publishing fields: Science, Technology, Medicine (STM), Business, and Transport.
- Springer eBook Collection with over 160,000 titles available on link.springer.com.
- Largest open access portfolio worldwide, with over 420 open access journals.
- More than 7,000 employees worldwide.

Key figures are Julius Springer—founder, Derk Haank—CEO, Martin Mos—COO, Ulrich Vest—CFO, Jan-Erik de Boer—EVP IT, Syed Hasan—President STM Global Academic & Government sales, Peter Hendriks—President STM Publishing,

Gregor Karolus—President Human Resources, Joachim Krieger—President Professional Businesses.

Timeline:

1842: Julius Springer opens book shop in Berlin and starts publishing soon after.

1889: Kluwer established in the Netherlands.

1946: Office in Heidelberg opened.

1964: Greater international focus and founding of Springer-Verlag New York.

1978: Founding of Kluwer Academic Publishers (KAP).

1999: Bertelsmann acquires the majority of Springer-Verlag, combines it with their Professional Information division, and names it: BertelsmannSpringer.

2003: The British financial investors Cinven and Candover acquire KAP and BertelsmannSpringer.

2004: Creation of Springer Science + Business Media through merger of BertelsmannSpringer with KAP. The result is one of the largest publishers worldwide in the STM field.

2005–2008: Acquisitions including Humana Press, Bohn Stafleu van Loghum and BioMed Central.

2009: Funds advised by EQT and GIC acquire Springer from Cinven and Candover.

2011: Springer acquires MPS/Adis from Wolters Kluwer. 2012: Springer acquires reference manager software Papers.

2013: Funds advised by BC Partners acquire Springer from EQT and GIC.

From the outset (since the beginning until mid-1900s), Springer was a commonplace private company. At this early development phase, the world was changing quickly because of industrialization. From political cartoons and compositions, the new Julius Springer Publishing House immediately went to the conventional science area. To keep the fast speed of the general public advancement, the following Springer's move was further extension toward the front-line technical subjects. They also focused on medicine and design-related fields.

To acquire control in Germany, another advanced move was portfolio extension; this demonstrated several strategies including further concentrating on medicine, technical science and design, and powerful acquisitions. This rich portfolio, centring on the best quality materials and utilizing advanced research methods, assisted sustainable organizational development in the crisis of the First World War.

The following 50 years after the Second World War, the third era of Springer, distributers added new procedures focusing on globalization and IT. The key idea was segregating duties and concentrating on natural sciences in Heidelberg and on general management in Berlin.

During the 1990s, the new emergency slammed the main driver being digital book production and distribution. While many renowned distributors were cutting incomes and even quitting, Springer responded nimbly to this new challenge. The reaction was moving the focal point toward digital improvements including smart innovations in best-in-class IT-intensive processes.

The other technique acquired from the previous development stages was segregation of duties. As such, they used resourcefulness and responsiveness to deal with this new crisis.

IGI Global

Settled in Hershey, Pennsylvania, USA, IGI Global is a main worldwide publisher focused on disseminating smart and innovative research results that enrich the information available in a number of areas. Closely collaborating with prominent experts from driving foundations, IGI Global spreads quality publications in over 350 topics and 11 knowledge areas. These distributions embrace over 100,000 industry-driving specialists around the world, guaranteeing that each title contains the most accurate and best quality coverage. IGI Global's founder is Dr. Mehdi Khosrow-Pour, DBA.

After celebrating 30 years of publishing success, IGI Global supplies a steadily growing assortment of reference books, journals, and a wide range of InfoSci-Databases of cutting-edge digital titles, covering the most pursued ideas on one stage. IGI Global's obligation to distribute the breakthrough research outcomes, remarkable assistance, and a positive attitude combined with a strive to exhibit the hot research areas and underrepresented ideas make a competitive advantage for this distinguished publishing company.

In this generally short way, IGI had a few changes—moving from specialized to multidisciplinary research, using the force of visionaries in these fields, carefully keeping moral code and quality guidelines, and implementing continuous innovative trends. Regarding inclusion, the methodology of IGI Global is more extensive and universal than that of Springer.

The startup in 1988 was small: the new organization published journals and books on management featuring a few specialized areas. In around 10 years, they added many more research domains.

Around the new millennium, the e-publishing era started. The IGI's response to this crisis was transforming their business to e-journal development in 1999 and publishing digital books on their proprietary unique InfoSci Online Database Platform in 2002. The activities to embrace global audience included launching new or rare product types such as reference books (1999), dictionaries (2006), and encyclopedias (2005).

Rather than building costly workplaces around the world, IGI's administration chose distributing its printed and digital products to over 40 wholesalers in more than 20 countries. The e-editions were also sold through important e-stores such as Amazon, Barnes and Noble, etc.

IGI Global's strategies and principles include:

- Close collaborations with researchers worldwide.
- Following high ethical practices.
- Meeting state-of-the-art standards.
- Establishing and following agile publishing processes.
- Being customer-centred.

To implement the above, they apply the following practices:

- Rigorous, high-quality peer review.
- Transparent publishing process.
- Fast and efficient dissemination of the research results worldwide.
- Focus on new trends and emerging research within prospective areas.

The key factor of IGI's success is addressing diversity by means of:

- Versatile subject areas.
- Different types of publications such as reference/authored/edited books, encyclopedias, casebooks, and journals.
- Versatile e-media such as mobile content, e-books, e-journals, online courses, and videos [25].

The other key factor of IGI's success is establishing trust by means of:

- Recommending through libraries, institutions, and colleagues.
- Using approved and high-quality reviewers.
- Following ethics codes including COPE (Committee on Publication Ethics).
- Meeting high-quality publication standards.
- Providing open access to certain kinds of publications.
- Providing specific access agreements to the authors such as private use, translations use, and non-profit education.

In its relatively short 30-year business history, the company managed to gain success. After surviving the digitalization crisis, it offers unique products and assists young and promising researchers. As an international group, IGI Global addresses diversity in many aspects. The primary reasons for their success are business process agility and customer focus. All these factors make the company crisis responsive and sustainably developing.

4.3   Crisis-Resistant Improvements

Introduction

Over the past 50 years, the life of people and companies has completely changed as technology has become its indispensable part. While some companies are losing money and trying to keep the usual order of things, others, using new technologies and practices, get ahead of their competitors, becoming market leaders. Over the years, many companies streamlined their processes; however, to operate efficiently and sustainably they must continually improve through the application of technical and administrative innovations.

Today, economics, formed and established many years before, is undergoing fundamental changes. The rapid development of the new digital revolution has affected all socio-economic institutions. As the Internet began to spread, the amount of information collected and processed began to increase rapidly. So, in the new economy, data has become a key asset [11]. Digital transformation poses significant challenges for large companies, affecting products, business processes, sale channels, and supply chains.

Digital transformation opens up opportunities for organizations to impact established business processes. As companies embark on a digital transformation journey, they can optimize operations, explore new business opportunities, expand reach, and gain insights into decision-making that improves customer experience. A well-designed optimization strategy can lead to tremendous cost savings.

However, for all the benefits that digital transformation brings, not all companies are ready to take it. To stay competitive and active market players, many firms have to fundamentally rethink and redesign their current business model and processes. Forming a strategy for harnessing digital transformation is still a challenge. The main reason for this is the insufficient maturity of the current business processes and the lack of the necessary skills and competencies [5].

Digital Transformation

Understanding digital transformation

Digital transformation is mission-critical for companies today. To survive in the market and remain competitive, organizations must adjust to the digital age and identify ways to optimize their business processes in order to reduce costs and increase revenues.

A key digital transformation driver is the rapid Internet expansion. As of January 2020, 59% of the world's population are Internet users, for Russia this figure is 81%. Along with this, the amount of information collected has tremendously increased. Processing this data plays an important role in business development. Drawing up an accurate portrait of the consumer allows businesses to become client-oriented by creating customized services and products. Thus, data has become the backbone of digital business economy. All business processes centre around them, and new business models and ecosystems are being formed, involving the interaction of economic agents in cyberspace [18].

As such, the spread of the Internet has affected more than just an increase in the amount of data collected. Along with this, the business transaction costs changed significantly. Now, much less resources are spent on finding information and making deals. The network revolution has triggered platform-network business models [11]. In contrast to the traditional model, these focus on the development of relations between the consumer and the manufacturer, rather than to the production itself.

Microsoft states: "The main goal of digital transformation is to increase competitiveness, to enable the company to grow in an ever-changing economic environment. The difference between digital transformation and conventional automation lies in the dramatic increase in efficiency. Therefore, the successful implementation of transformation in life, as a rule, leads to the creation of new businesses and business models" [15].

To remain competitive in the digital economy, traditional companies must produce clear development strategies. Innovative technologies can increase business agility, improve service quality, and identify new business opportunities. Successful examples of old-fashioned enterprise digitalization reveal two types of such strategies: (i) improving customer relations, and (ii) improving processes through technology innovations [16].

Companies that adhere to the first strategy strive to increase customer loyalty and trust. Digital technology facilitates finding, ordering, paying for, and receiving services, and provides support anywhere, any time, and by any media such as email, mobile app, or social media account [26]. The basis of this strategy is the analysis of data collected from customers. Careful data processing allows better understanding client's requirements and offering personalized services.

In choosing the second strategy, companies use the collected data to understand how digital technologies can improve the quality of their products and services. These can be significant changes in technology, hardware, or software. To understand which area can be optimized, it is necessary to conduct an in-depth analysis of current business processes. So, unlike the first strategy, when introducing new technologies, managers try to anticipate and get ahead of customers' needs rather than adapt to them.

Digital transformation is more often perceived by Russian companies as an opportunity to significantly reduce operating costs [24]. Thus, optimization of ineffective internal business processes comes to the fore. As such, customer experience transformation takes a back seat. Therefore, the second strategy, aimed at improving the processes by applying new technologies, is currently more relevant to Russian companies.

The need for digital transformation

Integration and leveraging of new digital technologies are among the biggest challenges companies have or will have to face. There is no single sector of the economy or business sector, the operation, and progress of which would not be affected by the development of new technologies.

For many companies, digital transformation changes both routine business processes and approaches to doing business. Their agile business model changes frequently. For example, Chevron oil company, directly unrelated to IT, has increased production efficiency by 30% by using artificial intelligence to identify the best location to drill new wells based on historical data analysis. Changes are underway in education, medicine, banking, and many other industries.

Digital transformation has become a top priority in company leadership programs. About 90% of business leaders in the US and UK expect IT and digital technologies to be a major contributor to the strategic development of their company over the next decade [4]. Faced with the challenge of digital transformation and the need to remain competitive in their industries, business leaders must design and implement strategies that address the digitalization impact and improve their operational efficiency. Otherwise, their clients migrate to competitors, since these provide faster or more convenient/reliable services that enable smart technologies. There are many recent examples of organizations that failed keeping up with the new digital reality, such as Kodak.

Digital Transformation within the company

For a company, digital transformation is a difficult stage that affects many, and often all, segments within an organization. Managers must simultaneously balance the exploration and exploitation of their firms' resources to achieve organizational flexibility. This tradeoff optimization is a prerequisite for the successful transformation of any business. Today, managers often lack clarity about the multiple options and elements that they must consider in their digital transformation activities. As a consequence, they run the risk of overlooking important elements of digitalization or ignoring decisions that are more favourable to their firms' specific situations, which could have unintended adverse consequences. Recently, many large companies opened Chief Digital Transformation Officer (CDTO) positions. These leaders of digital transformation take responsibility for the company digitalization, manage the processes and set the development pace. Also, in business development, it is important that the board of directors understands and supports the decisions to transform the company.

Requirements for successful small business digitalization

Corporations often have more financial leverage for digital transformation [7]. They can afford big budgets, contract external consultants to apply smart technologies, and hire the best talents to drive the transformation. For example, in 2009, Walt Disney appointed Sherrill Sandberg, a Facebook executive, to their company board.

At the same time, small businesses have certain advantages in digital transformation. First, it is the time and money spent on digital transformation. While in a large corporation decisions are made by the board of directors, approved by investors, people are found to lead the transformation process, and then go through many more steps, in a small business, decisions are made and approved tens and hundreds of times faster. In SME, the problem areas are detected faster, which also saves the budget. Secondly, with digital transformation, the speed of making and implementing decisions is important. With the support of the CEO, who is knowledgeable and motivated to optimize the processes and use smart technologies to improve the business profitability and competitiveness, SME can improve their performance fast and transform their business efficiently.

Which steps should the CDTO take to promote the company's digital transformation success?

First, the company's management team must discuss and approve the digital transformation concept. Further, they must arrange meetings with the head of each department, to clarify the reasons and company's demand in these changes. At this point, commitment of each employee is critically important as their motivation triggers the transformation. Secondly, a detailed strategy must be developed to address the business transformation impact on the company's business processes, identify short and long-term benefits, and understand how the new methods constrain the current business processes. Next, the business allocates a budget for the transformation, conducts competitor analysis, and detects the mission-critical industry technologies to digitalize.

Cases of digital transformation of large companies

Digitalization in manufacturing: Uniqlo

Uniqlo is a Japanese casual clothing retailer. The company has over 2,196 stores worldwide. Uniqlo has a market capitalization of over USD 57.7 billion and employs over 52,000 people worldwide. In 2017, the company had revenues of USD 20 billion and profits of USD 1.5 billion. According to statistics, every fourth Japanese owns down jackets from Uniqlo. Forbes magazine reports that Uniqlo has a brand value of USD 8.6 billion and is ranked 84th on the world's most valuable brands list. This is largely due to the founder's innovative strategy and customer focus [3].

The company develops every year; in 2018, Uniqlo introduced the option to buy, try on and immediately return to their physical stores the items that the customers did not purchase. Thus, Uniqlo changed the usual business model by adding e-commerce to its physical stores, thereby increasing its audience, and adding loyalty to old customers.

Uniqlo intensively uses smart technologies to improve and optimize its internal processes. As such, in 2019, their largest Tokyo warehouse replaced 90% of the employees with robots. These robots do many warehouse operations: they receive clothes, sort them, send them to other stores, and collect order packages. This eliminates "human factor" errors and increases warehouse operation speed and efficiency. Perhaps in the next 10 years, the company will start saving money by hiring no more employees. However, logistics speed and quality are the current Uniqlo top priorities. The company reported plans to invest over USD 850 million in smart automation in the near future. Uniqlo believes that this new digital infrastructure will lower storage costs, deliver products faster worldwide, and, importantly, foster sustainable development.

Ignoring digital transformation: Eastman Kodak

Eastman Kodak was founded in New York in 1892 by George Eastman, who created film cameras in 1888. He developed a business model in which families could buy a simple, inexpensive camera and shoot with film, which then needed to be developed. At the time, it was a huge change in the way we approach photography, transforming this from a professional activity into something everyone could do at home.

With the advent of new digital cameras, this photographic empire fell. Digital cameras gained popularity in the 1990s. Their introduction has revolutionized photography like the one that Kodak once created. Initially, Kodak was confident that customers would continue to see tremendous value in high-resolution printing, so they funded an aggressive advertising campaign to highlight the benefits of film quality over digital. However, digital cameras have proven that for a huge part of the market, the flexibility that a digital camera provides (for example, the ability to take an infinite number of pictures without incurring any cost) is much more valuable than the high quality printed images from film cameras.

When these changes in the market became obvious and impossible to ignore, Kodak decided to bring their digital camera to the market, but it was too late. They quickly lost market share and found themselves totally uncompetitive. So, remaining forever a manufacturer of film cameras, Kodak left the market.

Surprisingly, Kodak could become the first manufacturer of digital cameras. The company's engineer Steve Seison told the New York Times that he invented such a camera back in the 1970s; however, the managers decided that this innovation was not interesting for the company, as it would deprive it of the opportunity to earn money from selling films. The managers were not ready to give up their usual sources of income and, as a result, they eventually lost their entire business.

Transformation in the Toy Industry: LEGO

The LEGO company appeared in 1932 in a small town in Denmark and was originally engaged in the production of wooden toys. In 1958, the first plastic brick was released, which became the company's trademark. The multi-coloured details of the constructor have opened up endless possibilities for imagination and creativity of both children and adults. So, from a small local enterprise, LEGO turned into a world-famous one. Like many traditional companies, to survive in the digital transformation, LEGO had to radically rethink its business processes.

In 2003, analysts predicted LEGO bankruptcy, as with sales of just over USD 1 billion, operating losses amounted to USD 228 million. In an attempt to save the dying company, its owner, Kjell Kirk Christiansen, invested a significant part of his personal funds in it and resigned as CEO, handing it over to Jorgen Vig Knudstorp.

Before taking the company to a new level, Knudstorp had to save the company from the crisis. Taking a management position, he was able to highlight problem areas. At LEGO, innovation was encouraged, although its contribution was not always adequately evaluated. For example, creating a new colour for construction kit parts that only slightly differed from the old ones increased costs significantly, but did not represent real value.

Another bottleneck at that time was the supply chain. By taking advantage of the SAP infrastructure, the company was able to simplify the distribution process and reduce the number of suppliers. Thus, supply chain optimization was able to increase LEGO's revenue by 11%.

However, the company optimization did not end there. Emerging electronic toys such as the PlayStation and Xbox have become serious competitors in the battle for children's attention. LEGO had to think about how to get closer to their customers using new technologies, and digitalization opened up countless options for this.

The company appreciated the potential of using social networks to interact with end consumers. Currently, LEGO has user groups, fan pages, and interaction channels across multiple platforms. Constructor lovers are able to unite, communicate, share their assembled constructors, and follow the brand's news. Among the active users are not only children but also adults who still love building from small plastic bricks. On YouTube, the LEGO group has even become the second most popular brand, with almost 100% of its content generated by users.

The digital transformation has also impacted the LEGO products. Currently, their toys combine the physical and digital worlds. In 2011, the company started releasing mobile applications in which it is impossible to complete any level without putting together a constructor [1]. An example is LEGO FUSION, in which children use physical bricks in a digital game on mobile devices by scanning them with an app.

LEGO managed to create a powerful enterprise platform that prepared the company for the next development stage. However, the process of transition from a traditional company to a digital one is not yet complete and most of the changes are still ahead. While LEGO has made significant manufacturing improvements in recent years, management insists that the development of people, processes, and platforms should be an essential and continuous process.

ThyssenKrupp AG

ThyssenKrupp is one of the largest industrial enterprises in Europe, formed in 1999 with the merger of the German concerns Thyssen AG and Friedrich Krupp AG Hoesch-Krupp. The company includes various divisions dealing with the production of steel, automotive components, elevators, and escalators. ThyssenKrupp Elevator manufactures freight and passenger elevators for residential buildings, offices, airports, and other facilities. Apart from the sale, the company also deals with and maintains its elevators. Despite the group's predicament today, its experience is a good example of the use of digitalization.

With the advent and increase in the number of high-rise buildings, both business centres and residential buildings, the demand for high-performance elevators has increased. Along with this, the requirements for their quality and safety have also increased. The ThyssenKrupp was engaged in the maintenance of the installed elevators only, when necessary, which carried a significant risk for their users. The service itself being a very costly procedure, the company had to decide on improving their current business processes to retain their customers and reduce risks.

The solution to the maintenance problem was MAX, the elevator monitoring system. The use of digital technology allowed advanced detection of potential disruptions. ThyssenKrupp equipped their elevators with sensors that collected data on their condition and sent it to the support centre in real time. The data was analysed, and in case of a critical event, the service personnel were warned. Introducing the system significantly reduced maintenance costs and elevator downtime.

Using smart technologies and digital tools, ThyssenKrupp has managed to revolutionize the business processes supporting its products. The elevator monitoring system proved that data can become a critical asset, add value to consumers, and generate revenue. MAX significantly improved the quality of service, which affected the willingness of customers to pay more. Increased service requirement transparency benefited the stakeholders and improved their relationships.

Surprisingly, ThyssenKrupp experiences hard times today. Compared to 2018, losses increased from EUR 62 million euros to EUR 304 million. The company says its earnings before interest and taxes fell to EUR 802 million from EUR 1.44 billion in the previous fiscal year. To save themselves, the concern put up ThyssenKrupp Elevator for sale. The company found it difficult to compete with rapidly growing Chinese manufacturers. As a rule, the costs of Chinese companies are lower, which allows them to reduce prices and beat off customers from the German manufacturer.

Perhaps, a deeper analysis of business processes and their optimization through the use of new technologies would allow the enterprise to reduce its expenses and remain competitive [17]. Let us recall that there is room for improvement in any manufacturing process, making digitalization a continuous and imperative process. Who knows if ThyssenKrupp will be able to get out of the current crisis or become another traditional company that has not coped with the digitalization?

Transformation in the Video Rentals: Blockbuster LLC

Blockbuster is a former largest video rental network in the United States. The first store was opened in 1985 in Dallas by David Cook. The company quickly gained popularity. While other rental locations could only offer a couple of hundred films to their customers, Blockbuster offered several thousand. At its peak in 2004, Blockbuster had approximately 9,000 stores.

The business model for video rental is simple and attractive. Watching movies has always been one of the favourite evening entertainments [26]. As the prices for cassettes were quite high, a few clients could afford buying them on an ongoing basis. This problem was solved perfectly by the rental. People came to the store, picked out films, and rented them for a few days, paying much less. In the case of a late return, a fine was paid. Therefore, each cassette quickly paid off and began to make a profit for the distributor. Although DVDs replaced VHS shortly, this business model continued to work.

Blockbuster was quickly able to rise above other video rentals. Their innovative barcode system and computerized ordering allowed them to expand to 10,000 films per showroom when their competitors only offered 1,000. The first store was a success, and the company began to expand. In addition to films, music and videos began to be rented. Blockbuster showrooms began to open outside the US as well. The company also became the largest video distributor in the country. Therefore, it is not surprising that when in 2000 Reed Hastings approached John Antioch, then the CEO of Blockbuster, with an offer to buy the young Netflix company for USD 50 million, he was turned down. However, this was the fatal mistake of the largest video distributor.

There was a flaw in Blockbuster's business model that brought in a lot of revenue for the company, but which its customers strongly disliked: penalties for late return of the cassette. The founders of Netflix did not like that either. To compete, they started by shipping the DVD by mail. Saving retail space dramatically reduced expenses. This resulted in more attractive offers and abolishing the fines. However, Netflix's biggest competitive advantage was its new customer experience model. There was no need to pay for each movie separately as it became possible to subscribe and watch anytime.

Surprisingly, Antico did not believe in the success of this idea and its future. Being a many-year leader, his company neglected the imperative to adjust its strategy. When Antico realized that Netflix became a serious competitor, he suggested investing in the digital platform development to save Blockbuster. However, the management disagreed with his plans. As Jim Keyes took over as CEO in 2005, he immediately canceled the innovative changes, believing that they were detrimental to profitability.

In 2010, Blockbuster filed for bankruptcy. Currently, the only store remains open, which has turned into a tourist attraction.

It cannot be said that Netflix killed the largest video distributor. Blockbuster built what they considered a perfect workflow and did not see any point to improve it. Their management missed the critical moment of the entire industry transformation, and that was the end of the company. Blockbuster's case demonstrates that even a successful business requires continuous monitoring, optimizing, and implementing smart technologies to innovate, avoid crises and develop sustainably.

Conclusion

Rapid IT development has changed the entire concept of the economy. The obsolete business models used for many years quickly become outdated and ineffective. Businesses that did not realize this quit, unable to withstand the competition. This happens even to former industry giants, such as Blockbuster. Importantly, digital transformation is not about generating instant income. Instead, this is a lengthy process that requires thorough analysis, planning, and investment.

The aim of the section was developing recommendations to optimize the business processes, increase profits, reduce costs and remain competitive in the digital transformation. We examined the cases of digitalization success and failure in large companies. These included positive stories of LEGO and Uniqlo and negative experiences of Blockbuster and Eastman Kodak.

4.4 Conclusion

This chapter discussed the later transformations of even more dynamic and competitive environments. As previously, case method was applied as a part of the ITC framework to identify the potential crisis causes and effects, and outline crisis management through "by-example" approach. The case studies included publishing businesses and fast-food ventures. Of these, some were relatively long established, such as Springer and Dodo Pizza, whereas the others, such as Drinkit and IGI Global were relatively new and rapidly emerging startups. However, the common feature of all these companies was their essential IT dependency. We also analysed a few large-scale businesses including LEGO, Uniqlo, ThyssenKrupp, Blockbuster, and Eastman Kodak. The ITC framework application revealed that the volatile business environments examined may often generate crises in IT-intensive production processes. To manage these crises in a critical digitalization period, the strategy requires a carefully optimized balancing business, technical and human-related factors. The result is potentially successful transformation and further sustainable development at the organizational, business unit, and individual levels. However, human-related factors being an integral part of the ITC "three-legged stool" did not receive enough coverage yet. Therefore, we address them in the next chapters.

References

1. Andersen, P., & Ross, J. W. (2016). *Transforming the LEGO group for the digital econome.* MITSloan Management
2. Andal-Ancion, A., Cardwright, P. A., & Yip, G. S. (2003). The digital transformation of traditional Businesses. *MITSloan Management Review, 44*(4).
3. Baskin, J. S. (2013). The internet didn't kill Blockbuster, the company did it to itself [Electronic resource]. Forbes. Retrieved February 17, 2022, from https://www.forbes.com/sites/jonathans alembaskin/2013/11/08/the-internet-didnt-kill-blockbuster-the-company-did-it-to-itself/#74a a7e2f6488
4. Bonnet, D., Ferraris, P., Westerman, G., & McAfee, A. (2012). Talking 'bout a Revolution. *Digital Transformation Review, 2*(1), 17–33
5. Davis, T., & Higgins, J. (2013). A blockbuster failure: How an outdated business model destroyed a giant. College of Law Student Work. a. a.Lego wobbles as Star Wars and Harry Potter sales tumble. Retrieved from https://www.telegraph.co.uk/finance/2872600/ Lego-wobbles-as-Star-Wars-and-Harry-Potter-sales-tumble.html (Last Accessed February 17, 2022), b. b.Annual Report 2006, LEGO Group. Retrieved from http://cache.lego.com/dow nloads/aboutus/annualreport2006UK.pdf (Last Accessed February 17, 2022), c. c.Christina Warren (January 11, 2010). "LEGO Click: Building Blocks Meet Social Media," Mashable. Retrieved from http://mashable.com/2010/01/11/lego-click/ (Last Accessed February 17, 2022), d.Thyssenkrupp reports wider loss and sets new guidance. Retrieved from https://www. marketwatch.com/story/thyssenkrupp-loss-widens-and-sets-new-guidance-2019-11-21 (Last Accessed February 17, 2022), e.Thyssenkrupp reports wider loss and sets new guidance. Retrieved from https://www.marketwatch.com/story/thyssenkrupp-loss-widens-and-sets-new-guidance-2019-11-21 (Last Accessed February 17, 2022), f.A German Dynasty Sells Assets to Survive. Retrieved from https://www.bloomberg.com/news/articles/2019-11-13/steel-roy alty-no-more-thyssenkrupp-sells-itself-off-to-survive (Last Accessed February 17, 2022), g.A brief, illustrated history of Blockbuster, which is closing the last of its US stores. Retrieved from https://qz.com/144372/a-brief-illustrated-history-of-blockbuster-which-is-clo sing-the-last-of-its-us-stores/ (Last Accessed February 17, 2022)
6. Dodo Pizza CEO. Retrieved February 17, 2022, from https://sila-uma.ru/
7. Hammer M., & Champy, J. (1997). *Corporation reengineering: Manifesto of the revolution in business.* Per. from English - SPb .: Publishing house of St. Petersburg University. 332 p.
8. Instagram of Drinkit. Retrieved February 17, 2022, from https://www.instagram.com/drinki tcafe/
9. Instagram of founder of Drinkit. Retrieved February 17, 2022, from https://www.instagram. com/nncoffeeyou/
10. Kinzyabulatov, R. (2019). *Modeling of business processes.* Publishing solutions, 124 p.
11. Kochetkov, E. P. (2019). Digital transformation of the economy and technological revolutions: challenges for the current paradigm of management and anti-crisis management. *Strategic Decisions and Risk Management, 10*(4), 330–341
12. Matt, C., Hess, T., & Benlian, A. (2015). *Digital transformation strategies.* Springer Fachmedien Weisbaden.
13. Nikitina Nastya: Support is no longer needed by the small, support is needed more by the big. Retrieved February 17, 2022, from https://place.lemma.ru/interview/nastya-nikitina
14. Ovchinnikov, F., Founder of Dodo Pizza, launched a digital coffee shop. Will he be able to conquer the market? Retrieved February 17, 2022, from https://incrussia.ru/concoct/drinkit/
15. Prokhorov, A., & Konik, L. (2019). *Digital transformation. Analysis, trends, world experience.* AlliancePrint, 368p.
16. Sebastian, I., Mocker, M., Ross, J. W., Moloney, K. G., Beath, C., & Fonstand, N. O. (2017). How big old companies navigate digital transformation. *MIS Quarterly Executive.*
17. Steblyuk, I. Y. (2019). Business processes in the context of digital transformation. [Electronic resource]. *Economics and Business: Theory and Practice, 3–2.* Retrieved February 17, 2022, from https://cyberleninka.ru/article/n/biznes-protsessy-v-usloviyah-tsifrovoy-transformatsii

18. Schallmo, D., Williams, C. A., & Boardman, L. (2017). Digital transformation of business models - best practice, enablers, and roadmap. *International Journal of Innovation Management*, *21*(8), 17 (1740014)

19. The video about the first quartal of Drinkit. Retrieved February 17, 2022, from https://www.youtube.com/watch?v=iL4nlCuU_Jw

20. The sheet with financial information about the Drinkit. Retrieved February 17, 2022, from https://docs.google.com/spreadsheets/d/1MBw5-V0N3TLyn2yrDXLJTd9FXwQ417FffADX30ns_UQ/edit#gid=0

21. The site of the Drinkit. Retrieved February 17, 2022, from http://drinkit.ru/

22. The video about Dodo brands. Retrieved February 17, 2022, from https://www.youtube.com/watch?v=81Ur2uHcbb4

23. "There is a national product in Russia called shawarma": how the founder of "Dodo Pizza" plans to conquer the market of doners and coffee. Retrieved February 17, 2022, from https://www.forbes.ru/karera-i-svoy-biznes/412945-est-v-rossii-nacionalnyy-produkt-nazyvaetsya-shaurma-kak-osnovatel-dodo

24. The video about the Drinkit. Retrieved February 17, 2022, from https://www.instagram.com/tv/CEEQ0q5h0eu/?utm_source=ig_web_copy_link

25. Tarasov, I. V. (2019). Approaches to the formation of a strategic program for digital transformation of an enterprise [Electronic resource]. SRRM. No. 2. Retrieved February 17, 2022, from https://cyberleninka.ru/article/n/podhody-k-formirovaniyu-strategicheskoy-programmy-tsifrovoy-transformatsii-predpriyatiya

26. Vk account of the Drinkit in Vk. Retrieved February 17, 2022, from https://vk.com/drinkitcafe

Chapter 5
Taming Human Factors: Diversity in Digitalizing Multinationals

5.1 Diversity in Multinationals: Improving Performance and Decision-Making

5.1.1 Introduction

Since the industrial revolution, international trade and the structure of national economies have been constantly changing. The most relevant changes that have influenced creation of the global economy and globalization have started during the 1960s with expansion of US firms overseas to European markets, later on becoming Multinational Corporations (MNC) and changing the course of economic development. Further consolidation of trade in the international market and globalization were favourable for MNCs to grow into a supreme international force reaching the economic value of some countries' GDP and thus being able to influence some developing countries. According to the Bureau of Economic Analysis of US Department of State, MNCs accounted for over 50% of the USA export in 2010. In 2013, UNCTAD reported that: "Value chains administered in various ways by TNCs now account for 80% of the USD 20 trillion in trade each year".

The growing rate of globalization sets a favourable environment for further deepening of MNC influence on global markets. Furthermore, globalization and international trade forged strong economic relationships between national economies and intensified growing interdependencies. Financial crisis of 2009 has displayed outcomes that were suspected by the international community but still unresolved. Researchers attribute the major reasons for the financial crisis to MNC financial misbehaviours and failure of corporate governance controls [18]. However, these reasons caused the "economic domino" effect that has spread worldwide. Thus, substantial role of MNCs in global economics and its dominating influence suggested a further increase in control and monitoring in corporate governance. Consequently, in view of the board of the directors as a MNC's primary administration body, let us examine application of diversity effect on it as a possible crisis mitigation strategy.

S. V. Zykov, *IT Crisisology Casebook*, Smart Innovation, Systems and Technologies 300, https://doi.org/10.1007/978-981-19-2231-2_5

Another important argument for the importance of this research is an increasing role of MNC as a sole contributor of foreign direct investment (FDI) in the world. Balaz (2010) demonstrates that FDI has a positive influence on GDP growth of the host countries (e.g., European Quartet or V4 alliance) and transparency is considered an important factor in attracting foreign capital necessary for small countries to participate in international trade. Terjesen (2015) argues that transparency is influenced by female representation on boards, and this serves as a signal to public on high ethical grounds [48].

5.1.2 Definition of Diversity

Although many researchers offered a brief definition of diversity, not all of them clearly differentiated between functional and social diversity. Gonzales and Denisi (2009) identified diversity simply as "differences between individuals on any personal attributes that determine how people perceive one another" [19]. They did not differentiate between social and functional diversity, instead grouping all diversity traits into a single category. Richards and Kirby (1999) offered a constrained definition as they defined diversity only as differences in a limited subset of demographic characteristics including age, race-ethnicity, and gender. A more expansive definition was offered by Jehn and Bezrukova (2004) who, starting with a generic diversity definition similar to that by Gonzalez and Denisi (2009), further refined it by multiple guidelines [19, 25]. In particular, they clarified the issue of visibility of diversity, which was not enough discussed before [25]. This definition was among the more comprehensive ones. However, the most comprehensive definition of diversity identified social, information, and value diversity as separate terms and identified multiple factors in each of these subtypes [24]. This definition is perhaps the clearest and most useful in our research. The definition of diversity by Zanoni and Janssens (2004) was also very useful as it applied to the context of the organizations that they investigated. This definition also encompassed the idea of power and its relation to diversity, a factor that is noticeably missing in many formal academic sources. Given the relative importance of power in the conflict between social identities, this factor would be significant in the practice of organizational diversity management. Some researchers left certain key terms undefined or defined them ambiguously. For instance, Rynes and Rosen (1995) in their discussion of diversity training did not define this particular term and even diversity in general. Bunderson and Sutcliffe focused on functional diversity; however, they did not define functional diversity, in general, referring to sub-concepts instead [7]. Pitts (2010) also did not define diversity, in general, although did clearly define an overarching concept of diversity management. Østergaard, Timmermans, and Kristinsson identified the various types of diversity without providing any explicit definitions, although they did discuss the issue of cognitive differences based on social diversity [35, 45].

5.1.3 Review of Research on Diversity and Performance

Pioneering research by Chandler and Ansoff established the motivations for diversification and the general landscape of the diversified business [2, 10]. Wrigley (1970) refined and extended Chandler's studies by investigating the various options open to a diversifying firm [55]. Based on the works of Chandler, Wrigley, and others, Rumelt (1974, 1977) investigated the relationships among diversification strategy, organizational structure, and economic performance [43, 44]. Rumelt used four major and nine minor categories to characterize the diversification strategy of companies. The major categories were single business, dominant business, related business, and unrelated business. These categories provide a spectrum of diversification strategies from organizations that remained essentially undiversified to companies that diversified significantly into unrelated areas. Using statistical methods, Rumelt was able to relate diversification strategy to performance. The related diversification strategies—related-constrained and related-linked (e.g., General Foods and General Electric)—were found to outperform the other diversification strategies on the average. Relatedness was defined in terms of products, markets, and technology. The related-constrained businesses were found to be the highest performing on the average. In related-constrained companies, most component businesses are related to each other, whereas in related-linked organizations only one-to-one relationships are required. By contrast, the unrelated conglomerate strategy was found to be one of the lowest performing on the average. Later, Nathanson and Cassano conducted a statistical study of diversity and performance using a sample of 206 companies over the years 1973–1978 [34]. They developed a two-dimensional typology that included market diversity and product diversity for capturing diversification strategy that refines Rumelt's categories. They found that average returns declined as product diversity increased, while returns, in general, remained relatively steady as market diversity increased. However, they also found that size plays an important moderating role on the relationships. For both the market and product diversity, smaller businesses did well as compared to larger organizations in categories marked by no diversification and in categories of extremely high diversification. Larger companies did significantly better than smaller ones in the in-between categories characterized by intermediate levels of diversification.

Researchers are mostly uniform regarding positive influence of board independence on its performance [22, 27, 48]. Hermalin (1991) notes that there is no or financially minimal relation between composition of board and business performance; however, outside directors may have positive influence on board decisions [20]. Further, CEO interest is to have inside directors on their side and support their interests, whereas for outside directors the goal is to remain independent and pursue shareholder's goals. Thus, there is a struggle over the influence. Therefore, potentially, external directors should be a positive influence in monitoring the CEO. Moreover, in support of it outside directors will react strongly and promptly on poor company performance with dismissing the CEO and reject poor decisions of management rather than inside directors [21]. Rosenstein and Wyatt (1990) found that examining stock price change based on a new appointment of independent director showed 0.2% increase in stock price; however, results were not conclusive due to inability to single out stock value changed on external director appointment to all other announcements in favour of company performance [22, 42]. Mahadeo (2011) states that in developing countries independent directors hold rather monitoring positions, and therefore their influence on performance is minimal or cannot be assessed [28].

Diversity and its effects on the board of directors are widely discussed in the research. Concept of diversity in boards may be viewed as differences in education, background, industry, geography, gender, etc. Nekhili (2012) presented a comprehensive explanation of diversity focus—diversity is focusing on three attributes: variety, balance, and disparity. Therewith, variety stands for diversity of factors, balance—how many factors are distributed across the board, and disparity—differences of factors, such as male/female, etc.

Currently, the legislative and governmental organizations recognize the benefits of diversity as 16 national corporate governance codes encourage appointment of the female directors, and 14 countries mandate gender quotas [48]. Reviewing the diversification trend, the share of female directors on boards increased from 5.6% in 1990 to approximately 17% by 2014 [48].

5.1.4   The Role of Diversity

The main tenet of the knowledge-based view is that expensive expertise, based on a broad range of individual experiences, is the key to organizational functioning. Therefore, organizational financial performance depends on leveraging expertise to achieve competitive advantage via know-how that competitors cannot easily imitate or obtain [5]. Miller and Shamsie (1996) categorize such resources as the creative or collaborative skills that enhance a company's ability to both develop and market competitive products [29].

Racial and gender heterogeneity results in wider ranging perspectives, diverse types of information and ideas within an organization and results in superior problem-solving and decision-making [4, 13, 37, 41]. Individuals from diverse backgrounds bring the organization multiple perspectives for problem-solving and strategy formulation [24, 54]. For instance, researchers propose that women and racial minorities offer "requisite variety" in organizations and this variety leads to increased communication and performance benefits that are the result of creativity and improved decision-making [3, 30, 52, 53, 56]. Moreover, McLeod and Lobel (1992) found that groups with heterogeneous ethnic backgrounds produced higher quality ideas and generated a greater range of perspectives and alternatives in brainstorming tasks than did more homogeneous groups [33]. Also, Cox et al. (1991) found that ethnically diverse groups made choices reflecting different cultural orientations in a two-party prisoner's dilemma game [12, 13]. These ethnic and cultural perspectives may provide organizations with unique insights into distinct local niches as well as potential international markets.

In sum, diversity can promote creativity and improve decision-making, and hence lead to superior performance [16, 37, 40]. In the appropriate context, managerial racial and gender diversity can result not only in sociocultural benefits, but also financial returns [11, 16]. We offer Participatory Strategy Formulation (PSM) as a crucial process needed to forcefully unleash the positive benefits of managerial diversity, and therefore overcome development crises.

Researchers emphasize that for businesses to benefit from diversity, they must emphasize inclusiveness within the organization [16, 38, 39]. For example, Pless and Maak (2004) highlight how an organizational culture of inclusion, which allows people with different backgrounds, mindsets, and ways of thinking to work together, is critical to unleashing any potential "diversity advantage" [38]. In addition, Swann, Polzer, Seyle, and Ko (2004) describe how a group atmosphere that encourages freedom of expression facilitates openness among diverse members [47]. Also, Shore et al. (2011) highlight that practices, which promote sharing information, participation in expression, and decision-making, should be reflected in measures of inclusiveness [45]. We present PSM as a measure of inclusiveness that, we argue, facilitates positive interaction among diverse management members. We focus on these features as opposed to wider scale features (e.g., diversity climate and organizational culture) because PSM is a construct that was designed and validated specifically for studying the vertical and horizontal strategy making processes within the management corps [18].

Diversity can be a double-edged sword, increasing the opportunity for creativity as well as the likelihood of communication difficulties and misunderstandings [23]. According to the social identity perspective, racial and gender heterogeneity hinders group cohesiveness [9, 54]. However, research also states that those differences within groups can be bridged. In order for diverse groups to be effective and efficient, they must be able to avoid destructive fault lines so they can reach consensus regarding group decisions [26].

Participative processes serve as one mechanism that may allow organizations to leverage their creative variety emanating from diversity while preventing the negative consequences of potential fault lines. Research by van Knippenberg, De Dreu, and Homans (2004) emphasizes that in order for diversity to be advantageous there must be a participative process in place that allows for multiple viewpoints to be considered [51]. PSM processes not only foster collaboration and socialization across levels of management but also between or among diverse members. PSM should therefore facilitate knowledge sharing across diverse management groups, an important feature needed to truly benefit from diversity and obtain a "diversity advantage". Increased communication and collaboration among organizational members are the key to success of organizations by allowing them to utilize divergent technical, creative, or collaborative skills and improve decision-making effectiveness through the pooling and integration of group resources. Therefore, the pitfalls associated with diversity can be diminished and the benefits can be amplified when PSM processes exist.

Given the potential benefits and costs of workplace diversity, and the potential for participative strategic making processes to improve organizational financial performance, it is possible to predict that group heterogeneity alone may not be advantageous if an organization is unable to take advantage of the unique insight, judgement, experience, and know-how of women and minorities [17, 30, 46]. Participatory strategy formulation processes are possible ways to leverage workforce diversity and cultivate creativity, insight, and capabilities. Participatory strategy formulation incorporates broad ranges of perspectives, knowledge, values, and skills, giving groups better tools for effective decision-making even in complex environments [8].

5.1.5 Positive Effect of Diversity

A recent BCG study suggests that increasing the diversity of leadership teams leads to more and better innovation and improved financial performance. In both developing and developed economies, companies with above-average diversity on their leadership teams report a greater payoff from innovation and higher EBIT margins. Even more persuasive, companies can start generating gains with relatively small changes in the makeup of their senior teams.

For company leaders, this is a clear path to creating a more innovative organization. People with different backgrounds and experiences often see the same problem in different ways and come up with different solutions, increasing the odds that one of those solutions will be a hit. In a fast-changing business environment, such responsiveness leaves companies better positioned to adapt and increases crisis resistance.

Another takeaway found was a strong and statistically significant correlation between the diversity of management teams and overall innovation. Companies that reported above-average diversity on their management teams also stated innovation revenue that was 19% higher than that of companies with below-average leadership diversity—45% of total revenue versus just 26% (Fig. 5.1).

In other words, nearly half the revenue of companies with more diverse leadership comes from products and services launched in the past 3 years. In an increasingly dynamic business environment, this kind of turbocharged innovation means that these

Fig. 5.1 Impact of diverse leadership teams on innovation revenue (*Source* BCG, 2017)

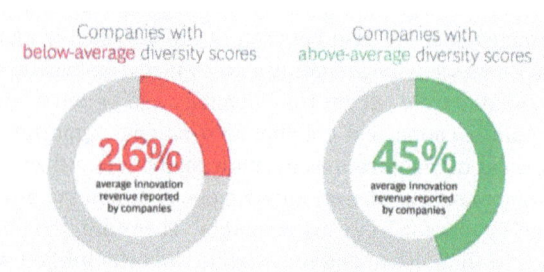

Fig. 5.2 Changes in leader and gain in innovation revenue (*Source* BCG, 2017)

companies are better able to quickly adapt to changes in customer demand, i.e., local crises.

To see how diversity on the leadership team can translate into better financial performance, they looked at a hypothetical company with about 50,000 employees and 1,500 people in management roles. They started with a diversity mix in line with overall averages and innovation revenue that was about one-third of the company's total (35%, the average for the dataset); they then changed each dimension in isolation to gauge the effect.

Relatively small changes in the makeup of management can have a significant impact. For example, if a hypothetical company was to hire 30 managers from a different industry (2% of the total management team), they would improve their innovation revenue by a full percentage point. Hiring 38 female managers (2.5% of the team) would have the same result, as would hiring 23 managers (1.5% of the team) from a country other than the one in which the company is based. To be clear, these are not incremental new hires but rather replacements for existing managers and executives; the overall size of the management team remains the same, but it is more diverse (Fig. 5.2).

This research also found that the effects are additive for all dimensions aside from industry background and career path, which show some overlap. Therefore, rather than focusing on any specific aspect of diversity, the goal should be to create teams that are diverse across multiple dimensions, as they all have value.

5.1.6 Research of Diversity Effects on Bank

In microfinance institutions, the results note that women are better able to lower operating costs and improve financial performance. In addition, studying this issue for Chinese companies over the past decade reports that business performance is positively related to gender diversity, although a critical mass of at least three is required for a positive effect.

The dataset contains 877 observations for 159 banks listed in nine countries: Canada, France, Germany, Italy, the Netherlands, Spain, Sweden, the UK, and the US. The period of analysis is 2004–2010, although at times there is no available information for some banks, i.e., the dataset is relatively unbalanced. The economic and financial data used to measure bank performance, bank assets, and loans were obtained from the CompStat database, while corporate governance data were drawn from two other databases, EIRIS, and the Spencer & Stuart Board Index (Fig. 5.3).

Explanatory Model 1a (Fig. 5.4) analyses the moderating effect that the level of investor protection has on board diversity. The results obtained show the same

Distribution of observations by year

TOTAL	2004	2005	2006	2007	2008	2009	2010
877	87	97	117	137	154	148	137
100%	9.92	11.06	13.34	15.62	17.56	16.88	15.62

Distribution of observations by country

TOTAL	Canada	France	Germany	Italy	Netherlands	Spain	Sweden	UK	US
877	67	19	23	66	25	56	21	186	414
100%	7.64	2.17	2.62	7.53	2.85	6.39	2.39	21.21	47.21

Fig. 5.3 Data distribution by year and country

	Model 1a		Model 1b	
	Coef. (Std. Err.)	t (p-value)	Coef. (Std. Err.)	t (p-value)
%WOMEN	74.874***	17.860	11.172***	3.90
	(4.192)	(0.000)	(2.867)	(0.000)
%FOREIGNERS	-3.748***	-4.850	-2.390***	-22.69
	(0.773)	(0.000)	(10.532)	(0.000)
%WOMEN*IP	28.282***	14.800		
	(1.911)	(0.000)		
%FOREIGNERS*IP	-1.343***	-2.630		
	(0.511)	(0.010)		
IP	Dropped			
%WOMEN*BR			9.821***	3.55
			(2.766)	(0.001)
%FOREIGNERS*BR			-3.056***	-24.55
			(12.449)	(0.000)

Fig. 5.4 Explanatory models

effect of the %WOMEN and %FOREIGNERS variables on Q as in the other model, regarding sign and significance. %WOMEN*IP presents a positive effect on Tobin's Q (coef. 28.282), statistically significant at the 99% confidence level, while %FOREIGNERS*IP has a negative and significant effect at 90% (coef. − 1.343). It can thus be affirmed that female directors in countries with lower levels of investor protection show a more limited impact on corporate performance (coef. %WOMEN = 74.874) than those who serve in environments with a greater concern for the defense of shareholder interests (coef. %WOMEN + coef. %WOMEN*IP = 74.874 + 28.282 = 103.156).

The results for models demonstrate a positive and significant effect at 99% of %WOMEN (coef. 11.172) and a negative effect at the same significance of %FOREIGNERS (coef. − 238.965). The interaction of both variables with the level of banking regulation (BR) supports these results. Specifically, %WOMEN*BR and %FOREIGNERS*BR have a positive and negative effect, respectively, both statistically significant at 99% (coef. 9.821 and coef. − 305.625, respectively).

The results demonstrate that female directors in countries with less stringent regulations have a more limited impact on bank performance (coef. %WOMEN = 11.172) than those who serve in environments with stronger banking regulation (coef. %WOMEN + coef. %WOMEN*BR = 11.172 + 9.821 = 20.993). In the case of foreign directors, their presence on boards negatively influences bank performance, especially in those environments with more stringent regulations for the banking sector (coef. %FOREIGNERS = − 238.965 vs. coef. %FOREIGNERS + coef. %FOREIGNERS*BR = − 2.390–3.056 = − 5.446).

Additional analysis

The later analysis reaffirms the strong business case for both gender diversity and ethnic and cultural diversity in corporate leadership and shows that this trend continues. The most diverse companies are now more likely than ever to outperform their less diverse peers on profitability.

The analysis of 2019 finds that companies in the top quartile for gender diversity on executive teams were 25% more likely to have above-average profitability than companies in the fourth quartile: up from 21% in 2017 to 15% in 2014 (Fig. 5.5).

While this approach is indicative, rather than conclusive, it could provide a more candid read on inclusion than internal employee-satisfaction surveys do and allows rapid and concurrent data analysis across dozens of companies. The focus is on three industries with the highest levels of executive-team diversity: financial services, technology, and health care. In these sectors, comments directly pertaining to inclusion and diversity (I&D) account for around one-third of total comments made, suggesting that this topic is high on employees' minds.

The comments analysed relate to five indicators. The first two—diverse representation and leadership accountability—are evidence of a systematic approach to I&D. The other three—equality, openness, and belonging—are core components of inclusion. For several of these indicators, the findings suggest the following "pain points" in the experience of employees:

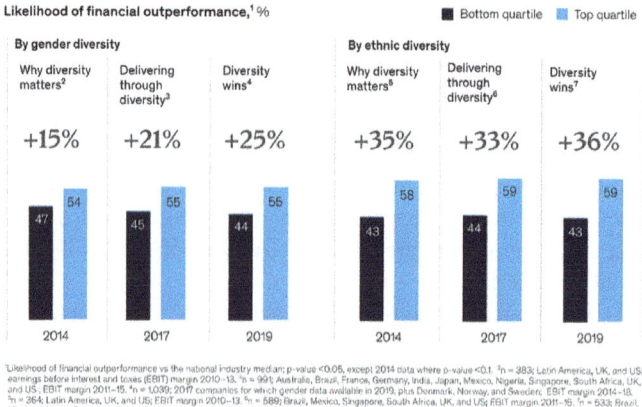

Fig. 5.5 Financial performance of executive diverse teams

- While overall sentiment on diversity is 52% positive and 31% negative, sentiment on inclusion is markedly worse, at only 29% positive and 61% negative. This encapsulates the challenge that even the more diverse companies still find inclusion challenging (Fig. 5.6). Hiring diverse talent is not enough; it is the workplace experience that shapes whether people remain and thrive.
- Opinions about leadership and accountability in I&D accounted for the highest number of mentions and were strongly negative. On average, across industries, 51% of the total mentions related to leadership, and 56% of those were negative. This finding underscores the increasingly recognized need for companies to improve their I&D engagement with core-business managers.
- For the three indicators of inclusion—equality, openness, and belonging—particularly high levels of negative sentiment about equality and fairness of opportunity were found. Negative sentiment about equality ranged 63% to 80% across the

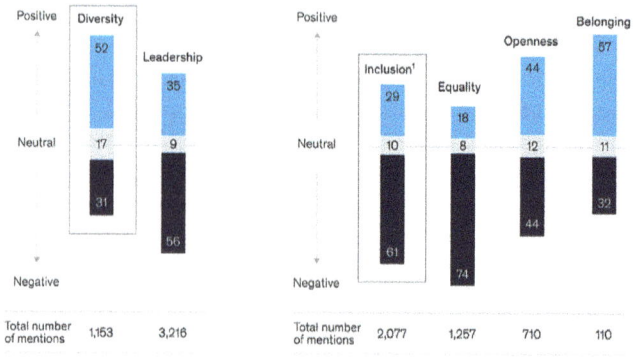

Fig. 5.6 Overall sentiment on diversity and inclusion, %

industries analysed. The working environment openness, which encompasses bias and discrimination, was also a significant concern as negative sentiment across industries ranged from 38 to 56%. Belonging elicited overall positive sentiment, although it had a relatively small number of mentions.

Let us take a close look at the more diverse companies, which are more likely to outperform financially. The common thread for these diversity leaders is a systematic approach and bold steps to strengthen inclusion. The best practices from these companies highlight five action areas (Fig. 5.6):

- Ensure the representation of diverse talents as this essentially drives inclusion. Companies should focus on advancing diverse talents into executive, management, technical, and board roles. They should carefully prioritize multivariate diversity options, such as going beyond gender and ethnicity, and set adequate data-driven targets to represent diverse talents.
- Strengthen leadership accountability and capabilities for I&D. Companies should place their core-business leaders and managers at the heart of their I&D efforts, i.e., beyond the HR or employee resource-group leaders. In addition, they should not only strengthen the inclusive-leadership capabilities of their managers and executives but also more emphatically hold all leaders to account for I&D progress.
- Enable equal opportunities through fairness and transparency. To advance towards a true meritocracy, it is critical that companies ensure a level playing field in advancement and opportunity. They should deploy analytics tools to show that promotions, pay processes, and the criteria behind them are transparent and fair; debias these processes; and strive to meet diversity targets in their long-term workforce plans.
- Promote openness and tackle microaggressions. Companies should uphold a zero-tolerance policy for discriminatory behaviour, such as bullying and harassment, and actively help managers and staff to identify and address microaggressions. They should also establish norms for open, welcoming behaviour and ask leaders and employees to assess each other on that standard.
- Foster belonging through unequivocal support for multivariate diversity. Companies should build a culture where all employees feel they can bring their whole selves to work. Managers should communicate and visibly embrace their commitment to multivariate forms of diversity, building a connection to a wide range of people and supporting employee resource groups to foster a sense of community and belonging. Companies should explicitly assess belonging in internal surveys.

One of the important practical implications of the above outcomes is that the decision to appoint foreign minorities to bank boards should be based on criteria other than the future financial performance of the organization. Nevertheless, the above evidence supports public policy initiatives for quotas of women on corporate boards based on the premise that gender diversity improves financial performance, especially in the organizations with high investor protection and strong bank regulation.

5.2 The Story of Microsoft

Introduction

In the recent years, Microsoft Corporation transformed. Currently, it is a rare public company with a market value of more than USD 1 trillion, and a top one by profit.

Nevertheless, Microsoft has significantly changed its strategy after the CEO change. As an established leader in the software and accessibility market, Microsoft seeks to implement the principle of universal accessibility in the field of innovative technologies. They continuously develop, promote, and introduce technologies optimized to meet the individual needs of customers.

The uniqueness of the Microsoft case is that the company had already built an organizational architecture for the implementation of the new corporate strategy by the restructuring program. This architecture helps the company not only advance the current restructuring program, but also create a venture that can adjust its strategy by creating new technological and organizational competencies. Thus, uncertainty of future technology solutions, efficient responding to competitor activities, and applying new opportunities became common features of Microsoft. The only formal company management entity is the new organizational architecture, which allows to survive in the crisis of changeable business environment and increasing technological complexity.

Satya Nadella, the Microsoft CEO, was honoured by the Financial Times as the Person of the Year 2019.

Walkthrough: Transformation at Microsoft

In recent years, Microsoft transformed, becoming a corporation with a market capitalization of over USD 1 trillion. How did Microsoft perform this change?

In Q1 of 2019, Microsoft reported net profit growth by 49%. The company promised that the remaining quarters of 2019 will show positive dynamics as well. In April, after the release of the previous financial report, the shares exceeded USD 137. This allowed the company's capitalization to exceed USD 1 trillion. Later, in June, Microsoft once again took a height of USD 1 trillion. This happened after a period of stagnation and even annual reputation loss due to massive product failures.

Founded by Bill Gates and Paul Allen, Microsoft enjoys a market value comparable to Apple, Alphabet, Facebook, and Amazon, the major tech giants of the decade (Fig. 5.7). In early 2000s, things were different with Steve Ballmer as the CEO. Steve succeeded Bill Gates being the second CEO in the corporation history, appointed in January 2000. During his 14-year reign, Microsoft gradually came from innovation to stagnation.

Steve's reign brought a few benefits: multiplied revenues, new software products, and a family of surface devices. However, along with these, Microsoft lost the battle for the smartphone market, the online search segment, and the leadership in cloud technologies [36]. In these areas, Apple, Google, and Amazon progressed better as they captured the new ideas and began to polish them, while Microsoft blindly sticked to their famous Windows and Office products.

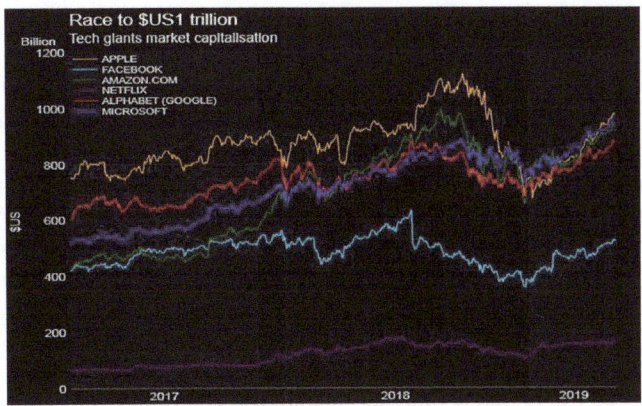

Fig. 5.7 Capitalization of the largest technology companies

Ballmer supported Microsoft isolation. When Apple launched iPhone, Microsoft CEO openly hated it, defiantly threw smartphones out the window when he saw the employees holding them and called open-source Linux operating systems a cancerous tumour.

In 2012, Balmer was "awarded" the "worst CEO" title of the USA public companies. Many employees were dissatisfied with his catching up management style, little vision and innovation, so that Microsoft lagged in the operating system and mobile service market.

In 2014, Balmer retired, and the new Microsoft CEO appointed was Satya Nadella, the former Head of Microsoft Cloud division (Fig. 5.8). Nadella became the leader of the transformation that resulted in Microsoft ultimate success.

The first thing Nadella did was putting aside all the legacy burden that many people at Microsoft trusted as the cornerstone of the company's business. "Our industry does not honor tradition. It only honors innovation", the new CEO wrote in his first email

Fig. 5.8 Satya Nadella, Microsoft CEO

addressing the employees. After this first appeal, Nadella started acting. What did he do?

The new CEO transformed Microsoft from a company centred around a single operating system to a business of diverse services and cloud-based technologies.

Nadella implemented unpopular decisions: massive employee firings and a few top manager displacements. He promoted cooperation between Microsoft departments. One of the main signs of these cooperative changes was the regular Microsoft hackathon where developers with very different backgrounds learned to solve problems together. As a result of this wise diversity management, many local crises of rivalry and competition between Microsoft departments remained in the past.

The new CEO developed positive attitude towards the market competitors. Unlike Balmer, Nadella promoted the principle of reasonable openness and even stated that Microsoft loves Linux, the competing operating system. Similarly, he released the Microsoft Office for iOS and Android platforms. At the presentation of the new Samsung flagship product, Galaxy Note, Nadella took the stage and announced a partnership with Samsung.

He conducted a campaign of aggressive promising project acquisitions, such as

- Minecraft game—USD 2.5 billion.
- GitHub IT service—USD 7.5 billion.
- LinkedIn social network—USD 26.2 billion.

Nadella restructured Microsoft to regain its status of the world IT leader. In 2018, he was awarded the title of the Best CEO by Comparably and the next year Financial Times awarded Nadella the status of the Person of the Year [6, 8].

The Digital Transformation: Takeaways and Pitfalls

Many people do not like Microsoft for its dark past and the decisions made by Gates, Ballmer, and even Nadella when they led the company over the years. Although Microsoft made many mistakes over the decades of its existence, unlike most competitors, it survived a number of crises.

At this point, many old Linux fans are reaching out for their guns but let us be honest: Microsoft has been giving less and less hate for the last decade. Windows 10 improves after each patch. Microsoft has a relatively small yet highly dedicated community of mobile platform fans. However, Satya Nadella's leadership is likely to reflect in the Microsoft history as the era of dreamers and visionaries. Let us try to explain the reasons for that.

The Epoch of Ballmer

Although Bill Gates started the Microsoft history, in terms of digitalization, the "era of Steve Ballmer" is worth taking a closer look at.

The period of Ballmer's leadership from 2000 to 2014 is worth mentioning at least for the fact that it was under his leadership that the company reached its "product" peak of strength and released two legendary systems, Windows XP and Windows 7. Under his leadership, the main work was carried out on creating Windows 10, the OS that still dominates the PC platform.

What is remarkable about the Ballmer era? First of all, the fact that Microsoft Ballmer is as far as possible from Microsoft of the 2020. The energetic, aggressive, and charismatic Ballmer is almost the exact opposite of the current Microsoft CEO Satya Nadella. Many people remember the legendary presentations with Ballmer and his shouts of "Developers! Developers!".

Steve took over the company at a dramatic time. Becoming CEO in January 2000 and replacing the founder of the company Bill Gates in this position, Ballmer got the company "on a horse": several months remained before the peak of the "dot-com bubble", the company's shares rose in price, and Windows XP project was in progress before its launch in August 2001. It may seem to many that under Ballmer, Microsoft, with their aggressive, unyielding policy, achieved success and rose in price annually. However, looking at the value of the company's securities during the Ballmer's leadership, one can deduce that Steve barely managed to keep Microsoft afloat after the crisis blow of the financial bubble to the industry giants in 2000–2001. Ballmer took over the company with a share price around USD 47 and handed in the keys to the office at that about USD 41. That is, indeed, for these 14 years, he did release something; however, he failed to develop the business as actively as it seemed from the outside. This is hard to believe, because it was the 2000s that were remembered by many as the era of total Microsoft product domination on all fronts: Windows XP, Windows Server 2003, Windows 7, and Windows Server 2008. MS Office, the de facto world standard of document processing software, was enhanced by Microsoft Office 365 that deployed web services based on the Office suite, and so on and so forth. Surprisingly, all this aggressive policy of capturing "the entire market at once" did not bring Microsoft significant profit, although it helped to stay afloat.

The New Era of Dreamers

Moreover, it was under the aggressive and "old-fashioned" Balmer that Microsoft began its transformation into a "dreamer" company. By the early 2010s, the company realized that it was impossible to drive further on a lame mare called Windows: the world was changing as platforms were becoming more portable and mobile, and everyday PC use was decreasing. It is not known what it costs the company's management to convince themselves and the shareholders that they again need to meddle in the mobile market. Let us remind that Microsoft were already pioneers in the field of tablet computers being well ahead of their time. We mean the tablet computer forgotten by many with x86 architecture, with full-fledged Windows XP control, a stylus and control buttons on the frame, which the company presented in 2003 as an alternative to laptops and PCs.

Many will say that the first notable step for Microsoft's return to the mobile market after the Tablet PC failure was their partnership with Nokia in 2011 as the Finnish company released the famous Nokia 800 smartphone running an early version of Windows Phone. Two years later, in 2013, Microsoft acquired the Finnish mobile corporation for a fantastic USD 5.44 billion in cash. However, this project has never succeeded even though the enterprise has been working on a tablet form factor since the 2000s.

Microsoft drew conclusions from the Tablet PC failure and began to develop their original idea of a working tool with a touchscreen. However, their conclusions were extremely specific, and instead of the iPad competitor, the engineers gave birth to the progenitor of the Surface product line, Microsoft PixelSense [31]. This touch-screen table was launched in 2007 and ran on Windows Vista. The main development purpose was an interactive showcase or info kiosk for shopping centres, banks, and museums. Both the Tablet PC in 2003 and PixelSense in 2007 looked as futuristic as possible. It is now impossible to surprise anyone with a huge touch panel as any bank or store has a similar info kiosk. However, 13 years ago, a desk-sized touchscreen display was a technology that even classic Star Trek did not have although its control panels were push-button. The creators of the series predicted tiny smartphones and tablets of the iPad format. By the way, this would not be a surprise if the Tablet PC were inspired by the same Star Trek.

Five years later, Microsoft returned to the consumer with a less revolutionary new product, and in October 2012, the Surface Hybrid Tablet PC ran Windows RT, the mobile version of Windows 8. According to the company, Surface was the direct heir to PixelSense in terms of technology. At that point, Microsoft engineers also did not follow the iPad fashion style to create a "pretty device for pointing a finger".

Instead, they wanted to make something useful, fantastic, and that would change the way millions of people think about the workflow. Again, a hybrid of a PC and a tablet was born: a work unit detachable from the keyboard base, combining touch-screen and external device input, and Windows 8 on board. Under Balmer, they released Windows 8.1-based generation for the Surface. The base models were no longer supported, although the Pro line devices still received updates for the next 6–7 years. So, what did Ballmer leave behind when he stepped down as CEO in 2014? Mostly a series of failures, such as Tablet PC and PixelSense. The reason, however, was that both projects were ahead of their time and were too ambitious. Tablet PC lacked the level of technology, whereas PixelSense was too daring and futuristic. Nokia's business also hung like a stone around Microsoft's neck: although they did an excellent job of creating a mobile platform, they failed to attract application developers to their ecosystem, which made it stillborn. Perhaps, the only relatively successful project that would have provided the company with a partial move away from Windows was the Surface, but Ballmer no longer developed that.

In 2014, he was replaced by Satya Nadella, the present CEO of Microsoft.

Satya Nadella and Microsoft's New Course

Before looking at Microsoft's transformation by the current CEO, it is worth exam-ining the company's financials. Ballmer just kept Microsoft afloat, while Nadella made shareholders richer at times. Taking over leadership at a stock price of around USD 41, the present-day CEO of Microsoft was able to increase this figure ninefold as the stocks of the technology giant exceeded USD 300 per share. What did Nadella do for Microsoft, and how did a stagnant company for a decade and a half suddenly become so attractive to investors? After all, Google, Amazon, Nvidia, Intel, and others have been growing all this time—about a dozen giants have increased their capital-ization and if their achievements were quite tangible, then what was Microsoft noted for? The first thing to remember is how Nadella started doing business as soon as he became CEO. Ballmer was a product of the 1990s: the hegemonic of Microsoft, he worked to establish the company as the undisputed leader and multi-area monopolist. Hence, the craving for revolutionary solutions, the dead joint project of smartphones with Nokia, the futuristic PixelSense, and the controversial Surface happened.

By the way, Lenovo engineers created their attractive Yoga Book, mimicking the Tablet PC and iPad gestures. Nadella, in fact, abandoned the total suppression policies and began to incorporate Microsoft into the existing framework of systems and services, realizing that the era of dotcoms and their dominance is gone, and negotiation becomes mission-critical. It is likely that Ballmer was very right chanting "Developers!" from the stage. The new CEO took advantage of this and went to the incredible—he started supporting the open-source community. Many people remember the moment when Microsoft officially became a Platinum Partner of the Linux Foundation, started supporting Linux and open-source developers, and even uploading their source code to GitHub, which shocked many developers [32].

By the time Microsoft was one of the main partners of the Linux Foundation, the company became the owner of the largest open-source repository on GitHub. And this is further evidence that Nadella has changed the domination strategy to inclusive interaction with the diverse partners and third-party developers. Therefore, in 2016, Microsoft made a progressive change, and Satya Nadella as a prominent visionary contributed to sustainable development of both the company and the entire industry.

In the next two years, Microsoft became the GitHub leader with the largest open-source section. In addition, the tech giant has released.NET Core 1.0 open source, worked with FreeBSD, and published a set of development tools. Microsoft currently partners with RedHat, SUSE, and other companies to develop open-source products. Nadella proved again that he was not only a successful CEO, but also a visionary in the company. "As a cloud platform for the company, we are committed to helping developers achieve more with their languages and platforms" said Scott Garty, Executive Vice President of Microsoft Cloud and Enterprise Group. "The Linux Foundation is home not only to Linux, but to many other more innovative open-source projects. We are thrilled to join the Linux Foundation and partner with the community to help developers capitalize on the ongoing journey to smart cloud and mobile".

And this was only the first bell. Taking a closer look at the company's activities, one notices that Windows 10 and newer OS versions have faded into the background for Microsoft. Five years after the release of the "dozen" (July 2015), no new versions are expected. Windows 10 is still the flagship OS from Microsoft, according to the company. Instead of global releases of their new products, the users receive continuous improvements. In future, this trend might result in an easier and more understandable licensed subscription for a couple of dollars a month instead of a lifetime license for a few hundred.

However, this is not just about new Windows releases. Microsoft has methodically sought out and supported many smart and innovative projects. In 2014, immediately after Nadella's arrival, the company spent USD 2.5 billion to acquire Mojang Studio, the Minecraft game developer, including all their patents. Few people remember that back in 2015, Ethereum, second most popular cryptocurrency after the Bitcoin, migrated to the Azure, Microsoft cloud platform, and is still there.

Many feared the GitHub purchase back then, because a number of tech giants usually have a harmful technique most clearly manifested in Google's policies: buy, destroy, close. Developers world over feared that Microsoft would start cutting GitHub's internal product tightly coupled with the Windows platform, or it would close it altogether putting an end to open-source partnership. However, the 2018 deal was relatively smooth. In any case, the prophecies that Microsoft will tie everything up on Azure and OneDrive, remove all open-source repositories, introduce a universal language based on Visual Basic, and, finally, turn GitHub into the Bill Gates Gulag did not come true. Over the years, there have only been more visionary projects. While Google, Apple, Amazon, and other tech giants report annual increase in display clarity, sales, and productivity gains of 3% over the previous generation, Microsoft is doing their own thing. Yes, the company does a good job of presenting new Surface devices, but what else have been heard about the broad consumer segment? Nothing like that, only a new Windows Phone incarnation, based on Windows 10X, came for the Surface Neo tablets.

Still, the recent research activities at Microsoft were very extensive resulting in endless news feeds. At a glance, Microsoft Research, their research division, addresses three key areas: "Artificial Intelligence", "Systems", and "Theory".

The fourth area, "Rest", that aggregates the projects beyond the above three. Therefore, research is being carried out in the most advanced and "futuristic" directions, outlined in the 2000s. These include computer vision, graphics and sound, machine languages, neural networks, quantum computers, privacy and cryptography, environmental projects, health care, and social sciences.

Conclusion

Microsoft Research website reports a complete list of their current activities. Customers do not just praise Microsoft, in general, and Nadella, in particular. Instead of battling Google in the browser and search-engine market or acquiring ARM to produce yet another type of CPU, Microsoft has successfully established itself in the B2B segment, squeezing money out of more mundane businesses around the world, and investing it in the future. In one of their presentations, Microsoft publicly almost shot their recent Internet Explorer and Edge browsers, opting for the Chromium engine. Almost, as they had to keep them in the Enterprise version to ensure backward compatibility with the large-scale legacy systems. This is the heritage of the Ballmer era. Holographic calls? There is such a project. Underwater data centres? Recently, they reported on the successful completion of the tests off the coast of Scotland [50].

Multi-criteria optimization of resource consumption for neural networking, object recognition, domain-specific language models to analyse healthcare research publications, formal modelling of the mankind behaviour—these are a few ongoing projects of the company, which initially used to trade in a proprietary OS. Microsoft has long outgrown the pants of the "software company" and is a huge ecosystem that sustainably implements the digital future.

5.3 The Sage of Huawei

Walkthrough

A man aged around 50 went out of an airport arrival hall after travelling in economy class. He ordered a taxi to a three-star hotel; this was an ordinary taxi rather than a luxury limousine. That was Ren Zheng Fei, the CEO of the Huawei Corporation. The emperor of this great MNC, he was a retired army officer and a former Chinese Communist party member.

The Huawei principles

1. Openness
2. Greyness
3. Compromise.

The second principle means shades of grey, i.e., that 100% black or pure white objects do not exist in reality. Therefore, one should approach any object as if it contains both black and white, i.e., positive and negative aspects. This approach is perhaps based on the Chinese Ying Yang philosophy also known as dualistic monism, where any black object contains a bit of white, and vice versa (see Fig. 5.9).

Following these principles, the early steps Ren Zheng Fei suggested were to avoid the dinosaur-like extinction of the fast-growing company in its early years. This fate was the history of such well-known and formerly famous multinationals as Deutsche Telecom, Alcatel, Motorola, and a number of other telecommunication companies. Let us see what made Huawei avoid these dangerous pitfalls which lead the above-mentioned multinationals to their collapse.

Crisis management and digitalization

The Sage book on Huawei says "In a sense, Huawei history a history of crisis management" [49]. Why is this so?

For the digitalization strategy, the Huawei CEO suggested the following steps:

Fig 5.9 The Yin-and-Yang symbol

(1)   establishing cloud services;
(2)   harnessing artificial intelligence;
(3)   keeping track of quality; and
(4)   following customer-focused approach.

Concerning the quality, a world famous Huawei's achievement was mounting the transmitters all the way up the Everest Mountain in the Nepal known as the highest peak in the world with nearly 8850 m top height. No other contractor ever tried to accomplish such a risky and challenging project. Huawei, however, did that in a relatively short term. In the atmosphere of a terrible, bitter coldness, harsh wind, oxygen deficit, and low pressure, the team managed to successfully mount all the equipment required. As a result, the climbers became able to communicate with their base camp and each other at any point of their route.

The Huawei story of development as a MNC is believed to be a multi-crisis history, the hampering factors being: (i) huge size and (ii) diversity in terms of culture, language, and a few other aspects. How did Huawei overcome these crises? Perhaps, the answer is in their ability to balance a number of mission-critical factors that affect the company development. The roots of this optimization are in the spiritual teaching which is known as Zen in Japanese, and Ch'an in Chinese, and which originates from the Indian tradition of Mahayana Buddhism. One basic principle of this teaching says: never overdo anything. One of later business management approaches based on Zen teachings and continuous saving is called Kaizen. The other well-known approaches of the kind include the strategy of "6S" optimization and the Capability Maturity Model (CMM). This requires multi-level, i.e., multicriteria optimization at the personal, business unit, company, and enterprise levels.

The other strategy for avoiding the crises was continuous improvement of the process and product development; this included the following:

(1)   developing the R&D department;
(2)   adopting best practices from the industry leaders, such as IBM Corp.;
(3)   bringing up the human resources, i.e., employees; and
(4)   using local traditions as the basis for the HR growth.

Knowledge management and agility

These practices are well aligned with the so-called "seven principles" for learning or acquiring knowledge [1]. These seven principles include the following:

• prior knowledge;
• knowledge organization;
• feedback and practice;
• mastery; and
• building a successful organization climate.

Following these principles means actively addressing diversity and involving the human factors, which allow for sustainable and crisis-resistant organizational development.

The essential principles introduced by the Huawei CEO were

- customer centricity,
- dedication,
- perseverance,
- openness,
- compromising, and
- greyness.

Once again, the latter item means that nothing is either pure black or 100% white. These business principles could be called the secrets of Huawei; however, they are tightly coupled with the above-mentioned "seven principles" of learning and knowledge acquisition. For example, the "mastery" principle is tightly coupled with continuous self-management which, according to the DMAIC process includes self-reflection, self-control, and self-criticism [15]. The latter principle serves as a vital source of novel idea generation and self-optimization at the personal, business unit, and enterprise levels. This continuous lifecycle of monitoring, analysis, and adjustment facilitates company progress and promotes crisis resistance.

The other potential source of crisis is diversity in culture, language, religious beliefs, etc. However, these aspects, if properly addressed, support the core values of the company culture and result in sustainable enterprise development as a heterogeneous team.

Huawei's keys to success

Surprisingly, Huawei addressed innovation in a rather traditional and therefore conservative way. In fact, the ingredients of their success include the following:

- gradual progress,
- conservatism, and
- stable growth.

These three essential ingredients lead the company towards sustainable development. The policy of Huawei is against any kind of radical changes, which are in fact a common cause of a crisis built artificially inside the company. This kind of crisis may appear very dangerous and even mission-critical in terms of the company survival and eventual business success. In contrast, Huawei suggests gradual progress in terms of process and product development. Several unsuccessful historical turnovers such as earlier Chinese dynasty-based revolutions and the recent cultural revolution resulted in dramatic lagging behind the other countries and tremendous efforts to catch up with them in the tough competition, including the IT-intensive businesses.

Ren Zheng Fei is often considered as the spiritual leader of the Huawei company. However, in order to reach this level of independence in decision-making he had to face an acute personal crisis. This happened when he was 44, which is considered synonymous to death as the words "death" and "four" sound alike in Chinese. He had to quit his former occupation, he was out of business and completely uncertain of his future. However, this personal crisis resulted in sustainable development of the company he created having barely USD 500 as a startup capital. The gradual improvement and critical thinking policy and following the "monitor-control-adjust" lifecycle made his company a prosperous and world leading MNC, while a number of seeming successful multinationals collapsed as mentioned above. Therefore, these two initially unfortunate digits of the 44 number had turned into a lucky product of 16, which in Chinese tradition means "a smooth rode ahead". As the Huawei story narrated, these crises, especially the events of 2005–2008, have "opened" Huawei, i.e., increased its agility.

During its development, the Huawei company faced a number of periods clearly associated with crises. To mitigate these crises, their CEO applied the strategy of "controlled democracy". This strategy followed the attitude that he inherited from the military, which, as put as a single word, could be represented as "obedience". This "obedience" phenomenon, formed by the military culture, contained the following key ingredients:

- Discipline,
- Uniform will,
- Team spirit, and
- Courage.

All these ingredients essentially correlate with the so-called agile approach to software development. Agility is a requirement to mitigate and eventually conquer crisis events.

The first decade that started 1987, while Huawei being a startup, featured extremely low budget, ambitious goals, and absolute authority. The following decade of 1998–2008 was characterized by opening triggered by a chain of crisis-like events. One of such notable critical events was the patent war between Huawei on the one hand and several multinationals on the other hand including the famous Cisco Corporation. Surprisingly, when the Cisco CEO visited the Huawei headquarters, around 400 R&D Department employees unanimously greeted him standing and applauding without any prior instructions. This shocked the CISCO CEO, as he encountered that the attitude of the Huawei employees was fair competition without ruining any relationships.

Another crisis happened in 2006, the reason being sudden death of Hu Xinyu, a 25-year-old Huawei's software engineer, followed by a trolling hype in the mass media. This event could be characterized as the crisis of a "false openness". Despite these and many other crises, Huawei demonstrated the unity and loyalty of the employees, and by harnessing the human factors the company eventually managed to overcome all these hardships.

The decade of 1998–2008 could be called the period of "quiet transformation". The idea was to refrain from radical changes in order to transform the company into an efficient large-scale MNC by means of a set of silent, reasonable, persistent, and firm activities. Around 2010, the Huawei Corporation became an even more agile organization, which was very competitive, conservative, and innovative together.

On their way to success, Huawei discovered a number of development process anti-patterns, i.e., dangerous pitfalls to avoid; these included perfection, appointing leaders without practical experience, and a few others.

Huawei considers the following phenomena as totally unacceptable while reforming:

- Perfectionism.
- Pedantry and minor nitpickings.
- "Blind" innovations.
- Partial optimization without improving overall performance.
- Reforming without a clear high-level plan, i.e., the "big picture".
- Appointing inexperienced reformers.
- Using any untested process.

Conclusion

As we can see from this brief case study, the secret of Huawei's success is a balanced combination of gradual improvement, self-discipline, self-management, self-criticism, and self-adjustment, which together with addressing diversity and harnessing human factors, based on establishing and maintaining corporate culture and traditions, improved the company's agility and crisis resilience. These vital ingredients, in a balanced proportion, paved the way to sustainable development of this smart and innovative large-scale yet agile enterprise.

5.4 Conclusion

This chapter explored the large-scale digitalization cases in diverse MNCs. First, general diversity aspects were discussed, based on a case study in multinational financial and banking organizations. The ultimate outcome was that management team diversity is a primary innovation driver, which, in its turn, if properly applied, drives digitalization and sustainable development. This finding has been widely exploited at large-scale MNCs such as Microsoft as an integral part of their MSF methodology or Microsoft Solution Framework. Further, this chapter presented a couple of case studies of crisis-resistant digital transformation of the above-mentioned Microsoft corporation, and the famous Chinese Huawei telecommunication giant. Microsoft faced several crises related to leadership also known as "CEO eras" of Bill Gates (1975–2000), Steve Ballmer (2000–2014), and Satya Nadella (since 2014) management. These crises originated from misbalance of the ITC "three-legged stool", i.e., business, technological, and human-related factors. Another case study examined Huawei, a corporation that although looks steadily growing and sustainable at the first sight, in fact, has evolved as it survived in a series of crises. These included crisis of growth, crisis of Internet technologies related to digitalization, and "crisis of openness". Surprisingly, since its foundation in 1987 and till the present day, Huawei is managed by one and the same CEO, Ren Zheng Fei. For these complex, diverse, and

IT-intensive multinational organizations, the key drivers for transformation strategies were identified. These drivers were based on innovative and smart technologies, amplified knowledge transfer, diversity management, informed decision-making, and agile client communication.

References

1. Ambrose, S., Bridges, M., DiPietro, M., Lovett, M., & Norman, M. (2010). *How learning works: Seven research-based principles for smart teaching.* Wiley and Sons.
2. Ansoff, H. I. (1965). *Corporate strategy: An analytical approach to business policy for growth and expansion.* McGraw Hill Book Co.
3. Ashby, R. (1956). *An introduction to cybernetics.* Chapman & Hall.
4. Bantel, K. A., & Jackson, S. E. (1989). Top management and innovations in banking: Does the composition of the top team make a difference? *Strategic Management Journal, 10*, 107–124.
5. Barney, J. B. (1991). Firm resources and sustained competitive advantage. *Journal of Management, 17*(1), 99–120.
6. Best CEO by Comparably. (2018). Retrieved February 17, 2022, from https://www.comparably.com/blog/best-ceos-2018/
7. Bunderson, J. S., & Sutcliffe, K. M. (2002). Comparing alternative conceptualizations of functional diversity in management teams: Process and performance effects. *Academy of Management Journal, 45*(5), 875–893.
8. Carmeli, A., Sheaffer, Z., & Halevi, M. Y. (2009). Does participatory decision-making in top management teams enhance decision effectiveness and firm performance? *Personnel Review, 38*, 696–714.
9. Carson, M. C., Mosley, D. C., & Boyar, S. L. (2004). Performance gains through diverse top team management. *Team Performance Management, 10*, 21–126.
10. Chandler, A. D. Jr. (1962). *Strategy and structure: chapters in the history of american enterprise.* MIT Press.
11. Cohen, G. L., & Garcia, J. (2008). Identity, belonging, and achievement a model, interventions, implications. *Current Directions in Psychological Science, 17*, 365–369.
12. Cox, S. F., Wall, V. J., Etheridge, M. A., & Potter, T. F. (1991). Deformational and metamorphic processes in the formation of mesothermal vein-hosted gold deposits—examples from the Lachlan fold belt in central Victoria, Australia. *Ore Geology Reviews, 6*, 391–423.
13. Cox, T. H., Lobel, S. A., & McLeod, P. L. (1991). Effects of ethnic group cultural differences on cooperative and competitive behavior on a group task. *Academy of Management Journal., 34*(4), 827–847.
14. Dess, G. G., Lumpkin, G. T., & Covin, J. G. (1997). Entrepreneurial strategy making and firm performance: Test of contigency and configurational models. *Strategic Management Journal,* 677–695.
15. DMAIC. Retrieved February 17, 2022, from https://www.dmaictools.com/
16. Dwyer, S., Richard, O. C., & Chadwick, K. (2003). Gender diversity in management and firm performance: The influence of growth orientation and organizational culture. *Journal of Business Research, 56*, 1009–1019.
17. Dwyer, J. R. (2003). A fundamental limit on electric fields in air. *Geophysical Research Letters, 30*(20), 2055. https://doi.org/10.1029/2003GL017781
18. Erkens, D., Hung, M., & Matos, P. (2012). Corporate governance in the 2007–2008 financial crisis: Evidence from financial institutions worldwide. *Journal of Corporate Finance, 18*, 389–411.
19. Gonzalez, J. A., & DeNisi, A. S. (2009). Cross-level effects of demography and diversity climate on organizational attachment and firm effectiveness. *Journal of Organizational Behavior, 30*, 21–40.

20. Hermalin, B., & Weisbach, M. (1991). The effects of board composition and direct incentives on firm performance. *Financial Management, 20*(4), 101–112.
21. Hermalin, B., et al. (1998). Endogenously chosen boards of directors and their monitoring of the CEO. *The American Economic Review, 88*(1), 96–118.
22. Hermalin, B., et al. (2003). Boards of directors as an endogenously determined instituition: A survey of the economic literature. *Economic Policy Review, 9*, 7–26.
23. Jehn, K. A. (1995). Amultimethod examination ofthebenefits and detriments ofintragroup conflict. *Administrative Science Quarterly, 40*, 256–282.
24. Jehn, K. A., Northcraft, G. B., & Neale, M. A. (1999). Why differences make a difference: A field study of diversity, conflict, and performance in workgroups. *Administrative Science Quarterly, 44*, 741–763.
25. Jehn, K. A., & Bezrukova, K. (2010). The faultline activation process and the effects of activated faultlines on coalition formation, conflict, and group outcomes. *Organizational Behavior and Human Decision Processes, 112*(1), 24–42. ISSN 0749-5978
26. Lau, D., & Murnighan, J. K. (1998). Demographic diversity and faultlines: The compositional dynamics of organizational groups. *Academy of Management Review, 23*, 325–340.
27. Macvay. (1999). The active board of directors and its effect on the performance of the large publicly traded corporation. *Journal of Applied Corporate Finance, 11*(4), 8–20.
28. Mahadeo, J., et al. (2012). Board composition and financial performance: Uncovering the effects of diversity in an emerging economy. *Journal of Business Ethics, 105*(3), 375–388.
29. Miller, D., & Shamsie, J. (1996). The resource-based view of the firm in two environments: The Hollywood film studios from 1936 to 1965. *Academy Management Journal, 39*(3), 519–543.
30. Milliken, F., & Martins, L. (1996). Searching for common threads: Understanding the multiple effects of diversity in organizational groups. *Academy of Management Review, 21*, 402–433.
31. Microsoft PixelSense. Retrieved February 17, 2022, from https://microsoft.fandom.com/ru/wiki/Microsoft_PixelSense
32. Microsoft is Platinum in the Linux Foundation. (2016). Retrieved February 17, 2022, from https://www.linuxfoundation.org/press-release/microsoft-fortifies-commitment-to-open-source-becomes-linux-foundation-platinum-member/
33. McLeod, P., & Lobel, S. (1992). The effects of ethnic diversity on idea generation in small groups. Paper presented at the annual meeting of the Academy of Management, Las Vegas.
34. Nathanson, D. A., & Cassano, J. S. (1982). Organization, diversity, and performance. *Wharton Magazine, 6*, 19–26.
35. Østergaard, C. R., Timmermans, B., & Kristinsson, K. (2011). Does a different view create something new? The effect of employee diversity on innovation. *Research Policy, 40*(3), 500–509
36. Richard, O. C., Kirby, S. L., & Chadwick, K. (2013). The impact of racial and gender diversity in management on financial performance: How participative strategy making features can unleash a diversity advantage. *The International Journal of Human Resource Management, 24*(13), 2571–2582.
37. Pelled, L. H., Eisenhardt, K. M., & Xin, K. R. (1999). Exploring the black box: An analysis of work group diversity, conflict, and performance. *Administrative Science Quarterly, 44*(1), 1–28.
38. Pless, N. M., & Maak, T. (2004). Building an inclusive diversity culture principles, processes and practice. *Journal of Business Ethics, 54*, 129–147.
39. Richard, O., Kochan, T., & McMillan-Capehart, A. (2002). The impact of visible diversity on organizational effectiveness: Disclosing the contents in pandora's black box. *Journal of Business and Management, 8*(3), 265–292.
40. Richard, O. C., & Shelor, R. M. (2002). Linking top management team age heterogeneity to firm performance: Juxtaposing two mid-range theories. *The International Journal of Human Resource Management, 13*(6), 958–974.
41. Richard, O. C., McMillan, A., Chadwick, K., & Dwyer, S. (2003). Employing an innovation strategy in racially diverse workforces: Effects on firm performance. *Group & Organization Management, 28*, 107–126.

42. Rosenstein, S., & Wyatt, J. J. (1990). Outside directors, board independence and shareholder welfare. *Journal of Financial Economics, 26*, 175–191.
43. Rumelt, R. P. (1974). *Strategy, structure, and economic performance.* Harvard University Press.
44. Rumelt, R. P. (1977). Diversification strategy and profitability. *Strategic Management Journal, 3*, 359–369
45. Shore, L. M., Randel, A. E., Chung, B. G., Dean, M. A., Ehrhart, K. H., & Singh, G. (2011). Inclusion and diversity in work groups: A review and model for future research. *Journal of Management, 37*, 1262–1289.
46. Shrader, Ch. B., Blackburn, V. B., & Iles, P. (1997). Women in management and firm financial performance: An exploratory study. *Journal of Managerial Issues, 9*, 355–372.
47. Swann, W., & B., Polzer, J., T., Seyle, D., C. & Ko, S., J. (2004). Finding value in diversity: Verification of personal and social self-views in diverse groups. *Academy of Managament Review, 29*(1), 9–27.
48. Terjesen, S., et al. (2015). Does the presence of independent and female directors impact firm performance? A multi-country study of board diversity. *Journal of Management & Governance, 20*(1), 1–37.
49. Tao, T., Chunbo, W., & Kremer, D. (2018). Huawei. Leadership, corporate culture, openness. Olymp-Business, 512
50. Underwater data centers. Retrieved February 17, 2022, from https://newsru-com.turbopages. org/newsru.com/s/hitech/15sep2020/natick_report.html
51. van Knippenberg, D., De Dreu, C. K. W., & Homan, A. C. (2004). Work group diversity and group performance: An integrative model and research agenda. *Journal of Applied Psychology, 89*, 1008–1022. https://doi.org/10.1037/0021-9010.89.6.1008
52. Watson, W. E., Kumar, K., & Michaelsen, L. (1993). Cultural diversity's impact on interaction processes and performance: Comparing homogeneous and diverse task groups. *Academy of Management Journal, 36*, 590–602.
53. Wiersema, M. F., & Bird, A. (1993). Organizational demography in Japanese firms: Group heterogeneity, individual dissimilarity, and top management team turnover. *Academy of Management Journal, 36*(5), 996–1025.
54. Williams, K. Y., & O'Reilly, C. A. (1998). Demography and diversity in organizations: A review of 40 years of research. *Research in Organizational Behavior, 20*, 77–140.
55. Wrigley, L. (1970). Divisional autonomy and diversification. DBA dissertation, Harvard University.
56. Zenger, T. R., & Lawrence, B. S. (1989). Organizational demography: The differential effects of age and tenure distributions on technical communication. *Academy of Management Journal, 3*(2), 353–376.

Chapter 6
Industry-Wide Case: The Russian Forest Industry

6.1 Digitalizing an Old Business

Today, the economy is going through worldwide changes because of the attention on advancement improvement. This is the time of new relations development dependent on advancements, both in the financial and social circles. Primarily, we discuss the institutional elements, which depend on financial advancement. Nonetheless, it is off base to talk about just broad institutional conditions for advancement without uncovering their points of interest for a specific industry.

The investigation of developments and institutional changes in the Forest business is especially pertinent because of the unique role of this industry in the public economy. Russia is the world's driving maker of wood. The possibilities for worldwide collaboration in the industry are growing. The environmental significance of the timberland assets identified with global climate change is becoming more acute. The backwood assets are diminishing in global utilization. The issues of forest fires, reclamation, and protection of different backwood areas are extremely poignant. The adjustments in the business are dictated by the national Forest Industry development strategy. The current over-bureaucratized institutional design of Forest industry did not permit the business to be efficient economically for a long time. The advancement of creative and innovative technologies was dramatically low. Over the course of the long stretches of changes, the executive's framework of the Russian forestry has undergone critical modifications, related to issues of proprietorship and legitimate administration, authoritative rebuilding, and timberland enactment. To implement the advancements and viable business in the Forest area, the entire rearrangement of timberland strategy is required, based on smart and innovative technologies. Institutional changes are developments affected by a number of variables. To date, the investigation of national institutional changes and the assessment of the viability of the associated advancements have not been adequately addressed.

To embrace the research subject, i.e., Forest business digitalization, let us centre around the idea of asset utilization and development, including eco-advancements. Eco-development is a shortcut for the term "environmental advancement"; this

© The Author(s), under exclusive license to Springer Nature Singapore Pte Ltd. 2022
S. V. Zykov, *IT Crisisology Casebook*, Smart Innovation, Systems
and Technologies 300, https://doi.org/10.1007/978-981-19-2231-2_6

includes issues associated with the environment, in relation to the approaches to their development.

In essence, the development in this area is related to prerequisites, such as the need to create efficient production and utilization cycles, which depend on ecological concerns. A variety of terms denote mitigating negative human effects on the ecology; these include eco-development, "clean" development, "green" development, economic advancement, environmental innovation, and "green" innovation.

The Organization for Economic Cooperation and Development characterizes eco-advancement as a movement that delivers an item or administration to recognize, forestall, limit, or fix harm to the climate, water, air, soil, and the issues related to environmental protection. It incorporates approaches and tools that decrease the risks related to the ecology, including climate changes, and protecting the environment from pollution.

From early industrialization till present, business has focused on profitability. From a business viewpoint, profitability fundamentally advances through upscaling that creates huge production volumes and decreases costs. Today, it becomes mission-critical to harmonize business efficiency with the capacity to analyse and implement innovations. The new IT advances changed the manner in which the teams cooperate. Organizations that ignore the execution of smart and innovative lifecycles in their headquarters decrease maintainability and become less client-responsive.

In view of digital innovation, the focus is on production process, advanced technologies, and administrative development. Product developments address the new "green" issues and help to decrease production costs. Progressive technologies change the design and innovation of making and using these products. Administrative advancements are related to organizational changes. These incorporate, for instance, updating organizational structure, expanding production volumes, implementing new production standards and novel production strategies, and redesigning the business. Such administrative innovations include intentional changes to improve the entire administrative structure, so that its parts synergize, i.e., accelerate, and operate more efficiently as a system.

Focusing on institutional changes, let us discuss the organizational types and attributes, such as:

1. The typology of establishments, depending upon their specialization, is either a "framework" or a "nearby authoritative". Frameworks are external establishments characterized by financial requests. Such establishments set the guidelines of financial operations. They incorporate financial principles as well as political and moral standards.

    The neighbourhood foundations are the establishments that include exchanges and authoritative organizations existing both in the open market and governmental structures. These are such organizations as stocks, banks, and business ventures. They are created to implement exchanges between diverse financial actors to decrease the levels of vulnerability and danger and reduce exchange costs.

2. The formal and casual organizations are recognized by the level of formalization of rules and guidelines. Yet, there is a conflict between the arrangement of formal

and casual foundations. Further, these split into (i) conventional foundations incorporating laws, sanctions, managerial principles, and guidelines, (ii) hierarchical and legitimate constructions, and (iii) casual organizations incorporating agreements, traditions, and moral and philosophical standards.

3. Social, financial, and political foundations are recognized by their activities under the idea of "institutional network". This idea implies a relatively stable arrangement of essential foundations managing the interrelated operation of such primary social subsystems as financial, political, and philosophical ones. The organizations of this kind guarantee the respectability of society. Such an institutional framework typically sets the limits, features, and directions of the financial processes.

The institutional change is identified and accompanied by the idea of institutional advancement. The institutional development means restructuring or rearranging expert collaborations or recombining the existing organizational components. This dramatically changes both the institutional status and its design, and results in a competitive advantage at a private, group, or public level. Therewith, such advantages being acquired at the individual employee level, as the research demonstrates, do not typically result in public advantages.

The institutional change is an advancement that unites both business experts and society in general. To trigger institutional changes, such advancements should meet the following prerequisites:

(i) degree of subjectivity and routineness of the changes, i.e., who, how, and in whose interests initiates these advancements.
(ii) degree of criticality of change consequences in terms of private, group, and public advantages.
(iii) likelihood of advancement rivalry.

In this respect, the institutional changes are important in view of their structural design. Typical instances of such development are companies and MNCs, including such aspects as their assets and possessions. Dayneko and Zykov demonstrate the influence of the company size on the dominating development directions for the entire business and specific projects [1].

Problems of the Russian Forest Industry and Institutional Improvement Framework

The Forest business is a significant part of the Russian economy; its condition impacts the Russian economy in multiple aspects, from the raw materials for a number of industries such as construction, agriculture, furniture production, and so on, to the timberlands as an environmental asset.

Russia is known as the world's biggest woodland country. Timberlands cover approximately 70% of its area and are among primary country's assets. According to the UN, Russia holds about 20.5% of the world's backwood territory and half of the coniferous woodlands. The majority of the backwood zones (29%) are the blended timberlands developed on the site of old cuttings, bunnies, and agrarian lands, similar to the Northern and mountainous forests. Another zone comprising 21% is covered by the larch woodlands. Coniferous timberlands cover roughly 19% of the woodland; these include 11% of cedar and fir woodlands, and 8% of pine backwoods. Leaf forests amount to 3%.

The backwoods area exceeds 1.1 billion hectares, which is 1/4 of the world's wood reserves. The Russian timberland woods cover around 75 billion m^3, which is many times more than that of the US or Europe. Over 20% of log and timber world trade comes from Russia.

The Russian woodland assets trigger the production of a few sectors of the national economy. The following enterprises deal with the Russian woodlands:

1. Ranger service takes care of the timberland assets and their utilization, prevention from fires, vermin, infections, and other negative environmental disasters and anthropogenic effects, saving and improving the regular habitat, assets, and ecological potential, and preserving biodiversity.
2. Wood industry system:

    a. Logging industry, which gathers and trades in various sorts of wood.
    b. Sawmill and wood processing industry, which produces timber, fiberboard, hardboard, chipboard, compressed wood, etc.
    c. Pulp and paper industry, which produces mash, paper, cardboard, paper packs, and similar materials, being a significant industry.
    d. Forest synthetic industry, performing compound handling and including dry refining of wood, carbonation, and different sorts of turpentine.

Wood processing industry is the fundamental wood producer. Russia trades in multiple sorts of timber items, i.e., roundwood, amble, fuel wood, sleepers, chipboard and fiberboard, wood, paper and cardboard holders, mash, paper, cardboard, backdrop, paper cleanliness items, etc.

The budget significance of woodlands is notable. Besides the financial advancement of society, the environmental role of woods is extremely important. The environmental woodland asset suggests administrative strategies such as world water and warmth, water insurance etc.; soil development and protection; biodiversity preservation; environment and climate changes; worldwide carbon cycle; and balneological and sporting jobs, among many others.

The role of backwoods is not limited by region they belong to. Russian woods are planetary important and play a significant part in the worldwide development of environmental strategies. Woods digest a huge portion of the world's carbon dioxide discharges, for instance, the woodlands of the Russian Irkutsk region amount to nearly 30% of the net carbon of the world's backwoods.

Other worldwide roles of the timberland are photocatalysis, carbon dioxide sequestration, and oxygen discharge. As such, according to rough estimates, the woods of the same Irkutsk region produce over 72 billion m^3 of oxygen yearly. Also, the backwoods of the Baikal Lake district provide water assurance, guide, and filter wastewater, and stabilize the water level. Backwoods' precipitous regions shield the soil from erosion, mudslides, snow torrential slides, and landslides.

Backwoods also clean, sterilize, and defend the environment from residue, sediment, and clamour. Wood shield guards from cold breezes and improves the environment. Backwoods decrease the impacts of destructive discharges. Preserving the woodland beneficially affects human wellbeing and innovativeness. Given this multiple worth of the woods, they are a vital national asset.

The key issues of the Russian Forest industry include the following aspects:

(i)   Innovative
(ii)   Environmental
(iii)   Institutional and governmental
(iv)   Insurance.

Creative and innovative issues include:

- Imperfection of logging and reforestation measures, low productivity of the backwoods framework due to a large number of logging regions over reforestation. In certain regions, timberlands gradually lose their environmental and water-managing importance. The reasons are coniferous backwood share reduction and a relative growth of the leaf estates due to inappropriate wood utilization. Reforestation is supported after self-cultivating, undergrowth obliteration, and soil corruption during logging and transportation of wood.
- Advanced processing and wasteful utilization of wood. Ideally, wood utilization suggests complete asset usage. In practice, the waste percentage is around 30% of the absolute wood yield. This happens because of fragmented logging, underlogging, and deserted trees. Advanced slicing methods lead to the reduction of wastes. Typically, bark (10% of the volume of wood), twigs (12%), and stumps (8%) are lost when gathering wood. Implementing innovative wood preparation methods reduces the wood deficit and, consequently, results in significant cutdown of backwood utilization.
- Inaccessibility of large backwood regions the development of which is challenging for current and prospective timberland clients.
- The absence of justified wood handling limits. To guarantee a better-balanced utilization of the common assets and generate higher profits, it is vital to improve the raw material processing. For a long time, Russia has been trading essentially roundwood, over 15 million m^3 yearly, which represents about 20% of the world's trading volume. The second product by volume is cellulose. Timber, paper, cardboard, pressed wood, and other wood handled items are sold in even less volumes.
- Low compatibility degree with the worldwide market, its principles and standards. The wood executives are to arrange a comprehensive woodland confirmation so that it becomes obligatory for stockpiling timberlands. Currently, the blackwood product line is very limited. Additionally, the Russian timber makers have restricted access to world business sectors.
- High level of wear and tear of the production lines. The deterioration of the machine tools in the forest production business in certain areas, such as the Irkutsk area, amounts to 80%. In the mostly capital-intensive pulp-and-paper industry, a

large percentage of the hardware is obsolete. At the best ventures, known as Scandinavian level, efficiency is around 30–40%, whereas at others it is about 10%. This challenge is additionally magnified by the absence of locally produced tools, and low involvement of private companies in their production.

Environmental factors include:

- The woodland sicknesses and bugs, to eliminate which it is recommended to utilize ranger services and systematic prevention activities.
- Forest fires being, perhaps, among the most significant challenges for Russia's woods. The root causes of such events are incorrect fire extinction attitudes and outdated technologies.
- Woodland ecological aspects are addressed by multiple timber organizations in Russia. Currently, the key ecological issue is incorrect backwood utilization; therefore, ranger services should be improved, and woods requiring reclamation should be reclaimed. Obsolete state legislation should be revised and updated. The inconsistency between the interests of timberland production's economic efficiency and wood preservation is another principal ecological issue. Managerial and legislative codes are the instruments addressing this issue to improve the Forest industry.
- Improving efficient wood processing in timberland regions.

The current organizational structure of the Forest industry hampers regional financial development together with the ability of Russian wood business to work efficiently. In its long development, the industry framework has gone through dramatic changes, the reasons being challenges in ownership, restructuring, and other updates in timberland production. The institutional design and changes in the backwood area are related to the strategies that direct the economic, social, and environmental development aspects of the woodland industry. There is an imperative for Forest industry innovations, developing federal and local strategies to support successful businesses, which requires an intensive institutional restructuring.

The fundamental institutional issues of the Russian backwood area are associated with woodland asset utilization and compliance with international ecological standards. Additionally, there are obvious legislative inconsistencies between the Forest industry and other branches of Russian laws. Various sources note the complexity of assessment strategies for the industry in its economic and management aspects, where multiple violations impede the business improvements.

The slow federal legislation progress hampers successful utilization of timberlands. However, recent advancements in Russian timberland laws have been announced as the new strategic outline. Specifically, this suggests improving Ranger service, integrating timberland protection and safety, and establishing state backwood control.

Tackling the issues of raw material shortage, consumption decrease, lack of properly certified and qualified staff, implementation of advanced wood production technologies, etc. are still vital. The industry progress depends on creative innovations, which require rebuilding institutional relations.

The issues mentioned show that careless timberland utilization occurs because of inadequate institutional relations in the industry, including improper granting rights to timberland assets.

The motivators for property policy improvement are:

- Establishing property rights granted by proprietor assurance from violations, including those originating from the state.
- Ensuring property privileges for the resources.

The following are essential property rights:

- Possession, including actual power of the property and renting.
- Removal, including the conditions and terms of use for asset owners.
- Utilization, including the privilege to individually own an asset.

In fact, before the approval of the Forest Code of the Russian Federation, there had been no proper legislation for timberland asset possession. This principal document stipulating the responsibility for these assets was issued in 1997. However, this Forest Code determined the privilege of ownership in a peculiar way. For instance, any asset considered as a federal type of woodland possession was controlled by the Russian governmental structures, whereas the privilege of its utilization went to backwood business structures as a lease. It is the Forest Code that clearly indicated possession rights for the timberland assets. The privilege of possession is completely determined when each privilege has its proprietor as the risk of property estrangement is mitigated.

Therewith, the property rights in the Forest Code were legitimized, although certain details were insufficient. Executive rights were split between the Russian Federation and its constituents; this often resulted in conflicts of interests. Money makers gained admittance to woodland assets, and timberland licenses; they maximized profits often being inconsiderate of public interests and ecology.

Whether or not timberland privatization was correct remains an open issue. Attempting to frame the Russian model of private wood possession ends up in comparing it to the models existing elsewhere. Privatizing Russian woodlands resulted in multiple negative social and ecological consequences. Russia never addressed private responsibility for the woodland ownership, nor had any clear terms and conditions for organizational and social obligations.

Obviously, the state influence on the improvement of the Forest industry looks insufficient. The state could and should better regulate the taxes paid through the collection framework, costs and compensations, and balance the market interest by establishing fair conditions for the development and restoration of the essential timberland products. The issues of the Forest industry are related not only to the market activities but also to efficient state guidelines that guarantee the harmony between economic efficiency and environmental protection.

Currently, Russian woodland laws and policies remain somewhat confusing. Forest service rangers repeatedly come across these issues when practically applying certain timberland guidelines. Besides, certain arrangements of woodland acts and policies conflict with the other environmental legislation acts concerning water, land, etc. For the industry, the draft of the new Forest Code of the Russian Federation did not resolve these issues. Progressing towards the market economy without a clear plan hampered the initial ideas of the Forest industry and Ranger service, causing a dramatic backdrop in the 1990s and mid-2000s.

Hence, creative changes in the Forest industry that can instantly boost business efficiency are probably unrealistic and should start with institutional changes in proprietorship and other related aspects. The essential condition for the Forest industry change on the inventive principle is the rearranged business framework.

Analysing the development frameworks centres around the factors and systematic measures, including different business landscapes, trading, and communication lifecycles. These lifecycles include the aspects of information, funding, and management. The experts analysed such frameworks in public development, local advancement, and economics. Of these, the first two kinds of frameworks are specifically significant to explore innovative development strategies in the woodland industry.

It is vital to shape up the business framework to advance it. Further, we use the concept of Forest Industry Innovation Framework (FIIF) meaning the institutional changes to develop the industry. This incorporates a set of interrelated aspects that make up a framework. The changes are controlled and restricted by the organizations that set the "rules of the game" in the Forest industry. This development framework is project based and determines a long-term strategy.

As the FIIF approach indicates, a "creative" business is typically beneficial and includes searching for better ways to establish enterprises, which improve administrative and development potential. An inventive framework should embrace the new strategies of the entire lifecycle and focus on product quality attributes, specifically, creating advanced versions of conventional products.

Such a business can be called inventive, as its lifecycle includes innovative developments and allows for new degrees of freedom, thus advancing this economic sector. A business venture is inventive when it permits a businessperson to advance by creating, producing, or trading innovative items.

In this way, inventive business is a mediator between logically formed business lifecycles as it promotes cutting-edge market trading. An important driver for inventive business is the societal imperative for cost reduction.

Inventive business allows for reducing expenses and minimizing intermediate costs. An optimized institutional structure guarantees efficient collaboration of business partners in the market economy. This is a fundamental part of the innovation framework as it applies new methods and opportunities for improving production in its financial, social, and environmental aspects.

Today, creative businesses should focus on the following measures to structure and advance the Forest industry:

(1)   Recognize the available resources of advancing the Forest industry based on its inventive potential.
(2)   Build the logging industry based on inventive advances.
(3)   Deliver production aimed at best quality.
(4)   Form buyer collaborations and extend markets.

(5) Improve production efficiency by advanced technologies and ecosystems.
(6) Evaluate and mitigate developmental risks.

The focus of the Forest industry institutional changes should be eco-friendly as it should improve production and administration lifecycles. As such, following the fundamental state guidelines should incorporate measures for institutionally transforming the businesses, including structural, administrative, authoritative, financial, and property aspects in their inventive transformation. The state instruments for managing industry development include financial and regulatory components. Their proportion is determined by the current financial circumstances and state planning guidelines.

An important objective of the innovation strategy is to expand the production level to guarantee the new product availability in the domestic and foreign business sectors, substitution of imported items, and sustainable development. Therewith, the principal goals of this innovative strategy are:

(1) Expand state investments in the research-based improvements, including environmental protection, safety, and networking.
(2) Initiate development activities in the public economy and social sector.
(3) Guarantee balanced coordination of cooperative international research in the national interests.

Despite the previous periods of changes, Forest industry relations in Russia have not formed a well-balanced framework. According to Nelson, the delicate process of institutional design should keep the communication "doors" open for previously available administrational frameworks [4].

The inventive adjustment of Forest industry policies in Russia should be based on an inclusive and innovative institutional climate for utilizing, securing, and developing the industry, and integrating it into the global community.

The institutional situation, where a certain proprietor is assigned to every key asset, meets the requirements for maintainability. The right of possession is legitimately fixed, and there is a procedure to secure property rights. The property rights for assets should be permanently fixed in the light of a legitimate concern for people in the future as these are fundamental for the society. The proprietor of certain assets should have a privilege to oversee and utilize these assets.

The institutional changes can either help or hinder the conditions for making innovations; the instances include:

- Contribute to adding value, including approaches to efficiently utilize assets, suggesting potential choices for their utilization, and redistributing the products without raising their rearrangement costs.
- Create value by introducing advanced methods of asset utilization and minimizing financial expenses after the rearrangement, following the planning and environmental protection standards.

The typical arrangement of financial adequacy indicators for innovative development incorporates:

- Aggregate sum of added value that requires analysis related to innovative technologies.
- Aggregate sum of added benefits, related to future development.
- Yearly profit from capital related to institutional development and its progress.
- Financial impact on the development based on net revenues and its dynamics in comparison with no improvement scenario.
- Financial impact on the innovation based on net yield/outcome, including deterioration and its dynamics in comparison with no improvement scenario.
- Capital restitution time of ventures focused on innovation and improvement.
- Aggregate sum of expenses according to the development budgeting plan in comparison with no improvement scenario.

For implementing, financing, and verifying the industry development, we suggest the following assessment factors:

- Innovation adequacy indicators: approvals for the development procedures, innovation-based deliverables, development difference value, i.e., gain/loss, investments related to development-based cost cutdown, and potential benefits resulting from development.
- Financial adequacy indicators: innovation-based compensation, net benefit, net gain, production volume, and production/utilization benefits.
- Venture feasibility indicators: innovation-based financial impact based on production, payment, and net benefits.

The attributes for institutional restructuring of the Forest industry should meet the following conditions:

- Ownership rights are clearly set and formulated.
- Rules are realistic and dynamically applicable.
- Decision-making is purposeful, transparent, and decentralized.
- Environmental and asset saving principles and laws are implementable.
- Private financial ventures are potentially profitable.
- Rule violators are traceable and punishable by law.

For a manageable Ranger service, the following criteria set for innovative institutional changes are recommended:

- Analysing the financial conditions for the business.
- Examining macroeconomic conditions to position the items on the financial "innovation map" of the country, its specific areas, and regions.
- Examining microeconomic conditions (e.g., for the Forest industry businesses and societies) including financial indicators, governmental and administrative organization, workforce, and innovation prospects.
- Institutional analysis, including a portrayal of the institutional construction of the Forest industry, financial business indicators, and institutional design assessment in view of the anticipated innovative changes.

The global goal is to implement the Forest industry innovation in terms of profitability and sociability. Further, let us discuss the impact of raw material supplies and their quality on the industry economics and social sphere in Russia.

At the microeconomic level, businesses and associations are the elements constituting the financial system. Therewith, the primary goal of the Forest industry development is addressing the issues for the sustainable timberland production and trade. For a long time, the State Ranger Service underestimated the role of local ranger services, associated organizations, business structures, and risk managers. Similarly, Russian central state agencies routinely neglected the contribution of regional lumberjack companies, mash and paper factories, wood processing plants, and sawmills.

We recommend an interdisciplinary methodology for the Forest industry development strategy to improve the Ranger service, incorporating the research in the following areas:

- Financial and social framework subsystems including ecological and innovative aspects.
- Inventive improvement.
- Institutional redesign.
- Analysis and assessment of the framework subsystems based on expert advice and quantitative strategies.

Deciding on the adequacy of the institutional infrastructure, we consider the following:

- Improvement rate and dynamics.
- Side effects of the changes due to incomplete/insufficient guidelines that result in "institutional traps".
- Analysis of the adequacy of institutional transformation including the impact of social inactivity and the need to keep distance from the conflicts, which often result from inadequate guidelines.
- Assessment of the improvement expenses by business experts.

The indicators for the institutional infrastructure adequacy in the Forest industry include the following:

- Medium salary rate for institutional experts.
- Income from the capital and other investments.
- Industry share in the GDP.
- Demographics of the experts on institutional infrastructure.
- Development rates.

Quality of the industry changes can be assessed by the success of achieving the goals and practical results obtained in a certain period. Therefore, in a crisis, the quality depends upon the ability of the actors to execute the critical changes reflected in the industry achievements. This can be implemented by simultaneous improving the outdated techniques and using the approved methods. Assessment of the improvements includes setting a timeframe and monitoring the resources required to achieve the goals and implement the policies.

Financial aspect of the institutional changes should be aligned with the social efficiency, which accounts for the ability of the people to implement and utilize the innovations. The indicators for these changes include profitability, product quality, and innovations as competitive advantages.

Let us use the following regression-based model to evaluate the innovations and institutional changes:

$$E_{if} = f\left(I_{tz},\ I_{to},\ I_{IT},\ I_{pro},\ I_{nano},\ I_{bio},\ I_{instform},\ I_{instinform},\ I_{org},\ I_{eco}\right),$$

where,

E_{if}	stands for the adequacy of the innovation of the Forest industry.
I_{tz}	stands for the mechanical innovations in logging.
I_{to}	stands for innovations in wood handling.
I_{IT}	stands for the IT-intensive innovations.
I_{pro}	stands for the product innovations.
I_{nano}	stands for the product nano-based innovations.
I_{bio}	stands for the product bio-innovations.
$I_{instform}$	stands for the institutional developments in the organizational foundations.
$I_{instinform}$	stands for the institutional innovations in casual foundations.
I_{org}	stands for the authoritative-based developments.
I_{eco}	stands for eco-based innovations.

The quantitative index of the industry innovation is selected to measure the sustainability of the industry improvement.

The above model meets the aforementioned prerequisites and is applicable to the analysis of the innovation impact on the Forest industry improvement. It allows for the following assessments:

- Innovation sustainability for a state or its certain area.
- Institutional development.
- Impact of a certain innovation element on the industry improvement.

The model for each element of the institutional developments is as follows:

$$D_{el} = f\left(N_{fed},\ N_{reg},\ G_{fed},\ G_{reg},\ I_{fin},\ I_{org},\ I_{prof},\ I_{info},\ I_{infinst}\right),$$

where,

N_{fed}	stands for the Federal administrative structure, e.g., for the governmental standards.
N_{reg}	stands for the local administrative arrangement.

G_{fed} stands for the actions of the federal government.

G_{reg} stands for the actions of the regional government including market orientation and sustainability.

I_{fin} stands for the financial area including agility and financial allocation practices.

I_{org} stands for the organization and management of the businesses and companies.

I_{prof} stands for the professional quality of the employees.

I_{info} stands for the data and information utility.

$I_{infinst}$ stands for the informal establishments including contracts and legally binding commitments, codes of morals and ethics, and local traditions.

Innovative Transformation Program for the Russian Forest Industry

The analysis of the current state and prospects of the Russian Forest industry allowed to determine the developmental effects on the production of wood. Economic and organizational improvements require an adjusted viewpoint for improving the human factors associated with the environmental changes, which is granted by the timberland maintenance policies [1].

The requirement for sustainable industry development becomes even more vital due to advances in economic globalization trends, which unite the wood businesses and expose an interrelated set of financial, social, and ecological issues. These include making high quality raw materials, establishing innovative production processes, and preserving and protecting the woodlands. Consequently, the key aspect of an advanced and sustainable Forest industry economy is the imperative for efficient Ranger service that centres around a balanced attitude towards the environmental, financial, and production issues.

Manageable improvement of the Forest industry includes financial, biological, and social maintainability, these aspects being tightly interrelated. Financial maintainability guarantees social supportability, and an eco-friendly industry would rely upon financial and social reliability. As such, the financial aspect of wood utilization includes the environmental element, and high ecological standards can positively contribute to the social and financial state of society. Social manageability indicates the balance among the above-mentioned supportability components.

In view of business improvement, we propose a the FIIF program for the Forest industry transformation incorporating four interrelated stages, which should be executed in the period of around six years, approximately 2021–2026.

Stage 1: Regulatory outline. This stage supports the State Wood Strategy, a priority in the Forest Industry Development Program. Radical changes in this strategy may prompt changes in recently adopted laws. In this context, the State Wood Strategy brings together the governmental experts and industry executives in Russia and globally, mirroring the social and financial framework designs of the country.

The issues of terminating the challenging and risky undertakings and efficiently searching for the leaders to manage the Forest industry ventures are the focus of the public authorities. The improvement plans include restructuring, reconstruction, re-profiling, and process innovation.

The primary components of the State Wood Strategy are improving the administrative structure, institutional systems, financial and economic conditions, human factors and HR, and social sector. The core of the industry strategy layout is its aim and objectives. This decision must depend on large-scale financial evaluation in the long term. The strategy development can be based on the share of backwood in the reserve, the proportion of the resources dispensed, resources to protect the woods from flames and bugs, etc.

This strategy is closely related to the industry administration; it contains the objective and major steps to improve the economy. According to the strategy, the industry administration positions its activities and projects in order to enable its improvement.

Stage 2: Optimization roadmap. An essential part of the Forest industry transformation at the State level is to decide on the steps to establish the resource optimization, which would assist the businesses and guarantee a proper balance between their profits and regional/Federal income.

To reasonably accomplish the industry improvement objectives, agile financial strategies, including duty and credit policies, this stage should promote mutually beneficial collaboration between State and business. For the industry clients, adequate lease and duty policies should be established and followed.

Stage 3: Business process optimization. The experts suggest the companies analyse their business processes, efficiently utilize raw materials, invest resources into business development, and extend their product offers. The recovery of the Russian Forest industry should begin with advanced wood production processes. There is a risk that the outdated production technologies, accompanied by financial odds and growing tax obligations on the roundwood, may result in raw material deficit and increase the product costs. As such, tax collection may hinder the business improvements, since the businesses meet a few challenges, including Ranger service-related issues.

To improve, the Forest industry companies should adjust their marketing, production, and raw material utilization strategies to meet the mission-critical factors. Specifically, it is important to analyse the target markets for each company, and to optimize raw materials delivery to avoid production crises. A key factor for the Russian companies is embracing the new Western and Asian markets.

Another critical factor is innovation efficiency and improvement. The current relatively low raw material, electricity, and manpower costs, tend to grow. Therefore, the partners should increase productivity to maintain reasonable costs.

The Russian wood processing companies need to improve their delivery cycles, enable effective expert involvement, and organize efficient ventures. Loggers need an advanced innovative cataloguing system of the wood items produced for the home and foreign business sectors.

Present-day industry adequately estimates production capacities; however, the exact product design breakdown requires special attention. The agile innovation progress sets new prerequisites that may critically affect the Forest industry and its economy. For instance, delivery option planning affects timberland production and processing. It is important to allocate the secured regions where mechanical production and processing should be prohibited. As such, it is critical to determine the asset zones for reasonable utilization where mechanical woodcutting is financially feasible. Since current activities in financially profitable backwoods are typically based on self-financing and the renters can contribute huge assets, it is radically significant to set environmental prerequisites for wood exploitation and reforestation. Additionally, non-wood plant life requires a set of preservation measures.

Stage 4: Financial Improvements. The measures required include the following:

- Financing sources should be justified, paying attention to woodland asset costs, based on leasing, and considering demand and supply volumes. These should include utilized asset reproduction, and special leases for excessive profits depending on the asset location and marketing conditions.
- Financial streams should be optimized.

Backwood production repayments should be set at the governmental level. Specifically, a part of these repayments should be returned to the timberland clients or kept in their accounting records.

Leasing compensation should be shared by several financial programs. For instance, the government should fund the timberland-related document processing costs and the allocated staff, including such procedures as backwood recovery and inspection, wood street development, and woodland security and protection.

Cost breakdown for the above measures requires an industry-level improvement plan. At the governmental and project levels, financing requires a clear and agile methodology.

Another essential problem is bookkeeping report improvement; this should be well aligned with the Forest industry development activities. Organizing and specifying the documented archives should accurately reflect the incomes and expenses of the timberland producers.

The financing should be based on a legitimate framework to ensure the reproduction of utilized timberland assets and guarantee their protection and preservation.

The improved subsidizing framework requires a set of adequate, long term, and well-aligned financial activities. Their justified distribution across the financial streams should balance the interests of the collaborating parties and provide a clear and reasonable financing strategy. Financial plans should meet the administration objectives and be manageable.

Conclusion

The problems of the forest business are intersectoral and interdisciplinary. The key prerequisite for the Forest industry development is, therefore, a systematic approach to backwoods assets and services, which includes continuous timberland regeneration. The activities should make up an interrelated set of measures that meet individual objectives of each timberland asset in a balanced way. This multi-criteria optimization ensures an aligned and efficient utilization of the timberland assets. As such, the woods are not only a national financial asset, but also the "lungs" of the planet. Therewith, preservation, regeneration, and innovative development of the forests require an integral nationwide program for timberland support by the Ranger Service and an effective Forest Code.

The industry issues to be addressed include innovation, environment, organization, safety, and reproduction. These issues call for thoughtful wood processing. The case study demonstrates that this requires restructuring and innovative development of the business and governmental structures of the Forest industry. The restructuring requires efficient collaboration of the institutions related to the industry. Thereby, any inconsistencies found should be properly aligned with the advancement goals and financial indices. The institutional innovations should improve the laws, rules, and standards of the Forest industry making a solid ground for effective woodland asset utilization. These goals include wasteless utilization of woods based on accurate mechanisms granting rights for their ownership.

A promising innovation is eco-development of the industry to decrease the negative effect of many businesses to the global climate.

The institutional developments should rearrange the components of existing business and governmental structures so that their synergy produces a nationwide competitive advantage. These institutional improvements should advance both financial outcomes and societal development. They should be balanced in terms of minimizing the negative impact and inter-departmental rivalry for the institutions involved.

The innovative framework for the Forest industry development incorporates disciplined management of assets, businesses, customers, technologies, analytics, and marketing, among other vital aspects. The core ingredient is the inventive business, which should incorporate the framework ensuring its sustainable development.

The new efficient Ranger Service should be based on an interdisciplinary approach, which incorporates financial and social subsystems including ecological aspects, innovative improvement activities, institutional redesign, and technology updates.

The institutional changes require investments and research. These institutional changes should be transparent to the public scrutiny. However, their practical implementation in law codes and business activities does not happen promptly, particularly in Russia, due to a high volume of participants and large territory. Therefore, these institutional changes require a thorough strategic analysis and quantitative assessment, particularly in the short and medium terms.

To assess the institutional changes, we suggest an integrated set of approaches including Institutional Analysis Development Framework, Political Risk Services (PRS), and the "Doing Business" assessment standards recommended by the World Bank for Development. Regression analysis should also assist in assessing the factors that influence Forest industry development.

To evaluate the innovative activities and institutional changes, we suggested a new model and applied it to the Forest industry of the Russian Irkutsk region. This model demonstrated the significance of institutional innovations. The experts argue that the industry's economic innovations rely mostly on causal factors including authoritative commitments and business customs. The innovative activities and institutional changes critically depend upon certain business parameters as prescribed by the model.

Public-and-private association is a promising institutional structure, where activities are supported by the reserves of the Government, Investment Store, Innovation Bank, and the local businesses investments.

Being many centuries old, the Russian timberland complex is still volatile in the market conditions, as it is hard to immediately gain agility in this industry's crisis. Our innovative venture analysis discovered that:

- Large-scale activities embracing the entire lifecycle are most efficient.
- Smaller activities involving innovative production equipment that make high quality items are also effective.
- Novel ventures and entrepreneurs quickly responding to challenges and increasing their labour efficiency also succeed in the crises.

The institutional changes suggested for the Forest industry are:

- Administrational improvement.
- Legal guidelines for woodland fires, including industry decision-making expertise at the federal level.
- State control of backwoods including fire monitoring.
- Improved law codes on timberland leasing.

Collaborative and efficient institutional relations are required to manage the intersectoral and interregional investments in wood processing and production facilities.

The proposed strategy for innovative development of the Forest industry suggests that it is mission-critical to support the projects of high-tech wood processing and producing items with high added value and innovation, such as biofuel. It is also important to support private companies implementing smart innovations and making new products. This strategy recommends State patronage for hi-tech innovations.

The financial support activities for the Forest industry include:

(1) Establishing resilient communication framework with logging industry.
(2) Separating administrative and technological management responsibilities to make new products.
(3) Innovating the timberland transportation to meet the best world standards.

The timberland producers should actively save the resources based on comprehensive leasing terms and conditions.

Transportation improvement, project innovation, and local/federal business development should be systematic and persistent. Along with woodland protection from unlawful activities, it is important to guarantee the rights for the business visionaries to legally diversify, innovate and contribute to sustainable industry development.

President Putin instructed the Russian government to stop uncontrolled unprocessed timber export and requested banning the shipment after December 2021 [5]. He also recommended subsidizing a state program to upgrade wood processing plants in 2021. "Undoubtedly, we need to act broadly in a number of areas. Exercising maximum decisiveness and taking into account the accumulated experience, [we'll] build an effective top-down chain of management in the timber industry. We need to engage in consistent, I would even say, a hard-hitting decriminalization of the industry", Putin said.

According to the Russian President, the timber industry is outdated. "It is necessary to improve the quality of state forest supervision. Overall, it is necessary to create conditions and a solid foundation for steady growth and development of the industry", he said. He also ordered the establishment of a federal state-owned corporation to coordinate the industry activities at the national level [2].

6.2 From Ad Hoc to Sustainable Development

Introduction

What is a quality software product? The intuitive answer is an error-free product. However, this ideal is almost unreachable in practice: in over 50 years of the software industry existence, developers did not learn how to create bug-free software. The search for methods and tools to get closer to this ideal is still vital. For example, the Capability Maturity Model (CMM) application steadily grows; this approach is widely used to measure the maturity of company development processes.

The quality management theory directly correlates the level of product development processes and the resulting product quality. CMM, designed to measure the development process maturity, currently dominates in Russia and world over.

The concept of Total Quality Management was formed in Japan in the 1950s, and later, in the 1980s, recognized by the Western digital production industry. It lays out the key principles that enable companies to build an organizational culture for quality IT product and service delivery. These quality management frameworks have been applied in a variety of quality models, of which the set of ISO 9000 standards was most widely used.

As ISO officially reported in 2003, over 500,000 companies worldwide are ISO 9000 certified. The advancement of these standards world over has played an important role in the organizational culture development for the businesses that seek to making quality products and delivering quality services. However, the ISO 9000 contains generalized quality management guidelines, and therefore requires a certain interpretation when it comes to company-specific implementation.

Milman, former Director of the Diasoft IT Company, said that the Russian software industry lacked strategy on focused CMM promotion.

Consumers usually need a certain level of software product quality, which calls for an industry-specific quality management model. The first customer of such a model was the USA Department of Defense; this happened as a result of systemic problems related to military software system maintenance.

Fig. 6.1 Quality triangle

Theory of quality management introduces the so-called "quality triangle" with the dimensions of resources, technologies, and processes. Figure 6.1 presents a view of this triangle in terms of scope, cost/budget, and time/schedule.

Achieving quality requires competent professionals and efficient tools. However, neither of the above two factors would give the desired result in the absence of a well-organized process, i.e., a sequence of steps leading to the implementation of the goal. Improving the skills of staff and using state-of-the-art technologies will not provide a breakthrough in quality unless efficient processes are put into service. The product quality is dependent on the process quality to create it.

Understanding this fact has become fundamental in creating the CMM model, which identifies the main groups of development processes, formulates the characteristics of different maturity levels of these processes, and recommends process improvement practices to reach a certain maturity level. The CMM develops an infrastructure for evaluating and certifying development companies to meet a required level of maturity. The impetus for certification processes was Pentagon's requirement that its software contractors and subcontractors require a formal maturity assessment of Level 3 or above [3].

In 2001, development of the CMM and its applications resulted in its upgrade to Capability Maturity Model Integration (CMMI).

Who needs the CMM/CMMI models?

The process approach to software development, formalized by the CMM/CMMI models, provides IT-intensive companies with a number of advantages. Quality processes allow organizing their operations to achieve the results desired, establish responsibilities, set key operation rules for the management, analyse operation progress, introduce quantitative parameters, optimize resource allocation, assess risks, etc.

Following the CMM/CMMI recommendations, companies build their process management and project planning, monitor and control projects, manage quality and configuration, choose technical solutions, and so on.

However, why only a few Russian developers implemented the principles of CMM/CMMI process models? To build such processes requires tremendous effort, time, and most importantly, investments. Dakhnovsky, the ex-Managing Director of the RUSSEE IT association engaged in consulting and training on CMMI, compared the preparation for CMM/CMMI certification with the implementation of heavyweight ERP systems.

Milman, being an expert in the CMM/CMMI practices, noted the two main reasons for encouraging companies to be evaluated to meet a certain level of model. He disagreed with the prevailing view that the company development costs increase if they have a certificate.

CMMI certification is required by the companies that plan to bring their products to the foreign, particularly USA, markets. Many USA customers would not even consider offers from developers without Level 2 CMM/CMMI Certificate. Large European clients also tend to apply similar qualification requirements.

Certification by a cooperative result of purposeful work to improve processes allows for fixing a certain state of processes in the enterprise to become a starting point for reaching the next maturity level and provides a firm justification for additional investments into further process development. According to RUSSEE, of the 2,500 CMM/CMMI certificates received worldwide in 1987–2005, around 50% were not advertised by their owners.

Before certification, a company typically sends a group of their employees to the SEI Official Introduction to CMMI training, which provides an overview of the models. Afterwards, they form an internal appraisal team to work together with external authorized assessors to certify in CMM/CMMI. Although this three-day training mitigates the crisis that usually arises as the development processes improve, it certainly does not solve all the related problems. There is a process named "Company Training" among the areas of the CMMI Level 3.

Implementing CMM/CMMI principles requires involvement of the most employees in training. Therefore, as Milman notes, this should be organized in several stages.

Further, the employees focus on different aspects of the CMM/CMMI model implementation, key concepts, and apply them to the current context of the company to get certified. The level of detail and discussion topics would vary depending on personal skills and responsibilities, though it typically requires the training of the managers and project teams, and the other employees involved in the key processes. The more individualized such training is, the more opportunities arise to compare the concepts and principles of CMMI with everyday practice, and the more productive the improvement is.

For the companies not ready for certification, it still makes sense to study the CMM/CMMI models. First, many of these model views may seem trivial, although it is often the obvious that is overlooked or ignored because of the usual human neglect and unwillingness to make changes in an already established process. For instance, training in compliance management highlights that reducing the software project requirement planning time often results in the need to update the system, and as a result, its development cycle is excessively lengthened causing extra project expenses, i.e., may result in a local crisis. Project planning courses teach documenting the development steps to ensure that the plan is realistic.

Studying CMMI develops the analytical skills required for development process assessment and quantification. This includes skills to evaluate error correction costs, exit criteria for software product testing, and other critical activities on the work schedule. Additionally, CMMI inspires a development culture that essentially improves the product quality.

Implementing the model

In Russia, CMM/CMMI training is somewhat spontaneous as only SEI authorized instructors are eligible to train. Therefore, the Russian developers highly valued the initial Moscow seminars by Mark Polk, the CMM v.1.0 creator. Still, the Russian software industry suffers from a lack of a focused strategy to promote CMMI ideas, which also negatively affects the state of learning in this area.

Certain experts argue that an expensive CMM/CMMI certification is still relevant for the Russian developers. Reportedly, for many Russian customers, the acronym CMMI is meaningless and, therefore, the developers often do not require such a certification. Among the reasons for certification is the prospect of orders from the Western customers.

There is also an opinion that the CMMI model, unlike its predecessor, is relevant for large-scale IT developers only. Nevertheless, CMMI as a learning subject considerably interests the development community and shapes up their corporate culture. The other trend of the Russian enterprises that cater "to the West" is that they show increasing interest in the domestic market. The area gradually becomes more competitive, while the market leaders are the vendors that already mastered the CMM/CMMI practices.

According to RUSSEE, as early as in 2005, seventeen Russian companies have already become CMM/CMMI certified, and four more were processing for certification. At that time, the country already had an experience of obtaining the highest levels applying both models, CMM and CMMI. Examples included Luxoft and Motorola, the Luxoft's CMMI Level 5 certification being the first one in Europe.

CMMI maturity levels

- Level 1, Initial: Processes are ad hoc, or informal (see Fig. 6.2).
- Level 2, Managed: Processes are defined and documented; however, focused on organizing a particular project, i.e., not standardized, and often vary across projects.

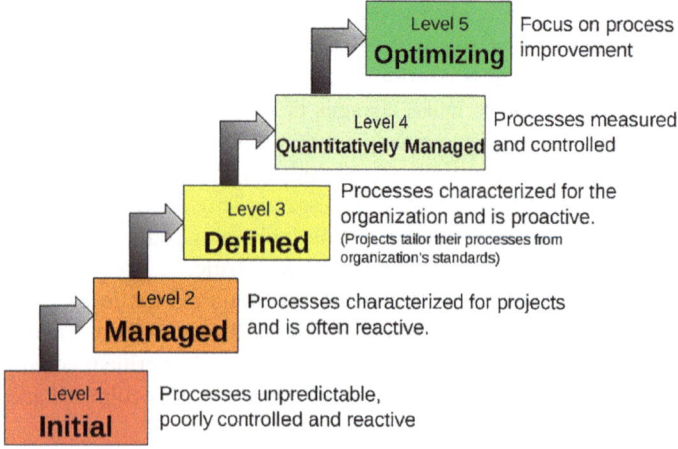

Fig. 6.2 CMMI levels (*Source* NASA, 2008)

- Level 3, Defined: Processes in all projects follow a preset corporate standard, i.e., the so-called company standard process.
- Level 4, Quantitatively Managed: Processes are predictable and manageable for such project parameters as error rate, labour, and processing volume.
- Level 5, Optimizing (Optimized): Processes are continuously improved; the company introduces significant innovations in their processes based on quantitative analysis, identifies the root causes of project issues, i.e., crises, and prevents such critical events.

The case of RUSSEE: Managing Software Development

In the early 1990s, the USA Institute for Software Research (ISR), in close collaboration with the Software Engineering Institute (SEI), developed a university master's program for software project managers in Russia, which incorporated the key ideas, concepts, and principles of the CMM model. Based on that, RUSSEE took one of its main subjects, Managing Software Development, and developed a concise version designed for on-the-job training, which included the following aspects:

- Managing technical requirements
- Planning and managing changes
- Quality control
- Risk management
- Project assessment
- Developer management.

Each session combined 50% lecture material and 50% practical exercises, including individual homework, case studies, and team assignments. Until recently, RUSSEE relied mainly on the Carnegie Mellon University teachers; however, later the Russian instructors were certified for all their training. The Requirement Management seminar delivery confirmed that for the Russian developers English was an adequate medium for communication.

According to Dakhnovsky, this program complied well with the CMM and CMMI standards, so that if a company implemented the recommendations of these seminars, they would typically get at least the Level 2 Certificate.

Development team leaders can choose workshops individually; taking all the six modules grants the students, and their companies, certificates from Carnegie Mellon University. After the first graduation in 2004, RUSSEE introduced assessment-based certification that allowed the graduates receiving official confirmation of a development team leader from Carnegie Mellon University.

RUSSEE intended to expand the range of subjects and in 2006, they adapted the university curriculum in software architecture. According to Dakhnovsky, the program was technologically focused and provider independent. The organization recognized that their add-ons to the higher education programs, based on the original Carnegie Mellon University curriculum, interested the Russian IT educators. However, implementing this program in the universities was a challenge due to a number of issues, such as the US licensing policy, and training the Russian teaching staff. Eventually, this was partially implemented at the Higher School of Economics as a program in software engineering. Additionally, RUSSEE managed to improve the training level of developers and process managers in a few Russian universities.

6.3 Conclusion

This chapter presented the case of Russian Forest industry digitalization. A major challenge for this significant yet very traditional industry was inflexibility of the organization and production processes. Clearly, this stiffness contradicted with the current state of fluctuating and unstable environment, subject to frequent technology and policy changes. This crisis threatening state was further impeded by a set of complexity factors that included organizational, legislative, and logistical problems. To manage this crisis fundamentally complicating the digital transformation of this important industry, agility improvement by tradeoff optimization strategy was proposed. Another direction of crisis resistance outlined was organizational process maturity. This domain was explored based on CMM/CMMI models, which typically have five levels starting at the ad hoc and ending at the optimized processes. Improving process quality and maturity level is the practice recommended to large-scale enterprises to achieve crisis resistance and develop sustainably.

References

1. Dayneko, A., Dayneko, D., & Zykov, S. Crisis of institutional change: Improving restoration and reconstruction methods for estate cultural heritage. Retrieved February 17, 2022, from https://www.semanticscholar.org/paper/Crisis-of-Institutional-Change%3A-Improving-and-for-Dayneko-Dayneko/40a53842b8b84d7e3274418a4ae3dc0244ce1937
2. Federal state-owned corporation to coordinate the industry activities at the national level. Retrieved from February 17, 2022, from RBC.ru https://www.rbc.ru/business/13/05/2021/609a92599a79473c9479bd75
3. Maturity Assessment by developers. (2005) *Open Systems & DBMS Journal*, (2)
4. Nelson, R. R., & Winter, S. J. (2002). Evolutionary theory of economic changes. Delo
5. President Putin instructed the Russian government to stop uncontrolled unprocessed timber export and requested banning the shipment after December 2021 (Sept 30th). Retrieved February 17, 2022, from https://tass.com/economy/1206747

Chapter 7
Social Aspect of Digitalization: The "Human Factors"

7.1 Knowledge Transfer: Digital Transformation

Introduction

In the current complex world, knowledge becomes a real asset that drives the business environment to progress. Knowledge allows us to innovate and improve climate, decrease spending, and propose new products to customers. The main sources of information can be market contracts, internal research, and relational agreements. The first type can provide information about a product and its technical specificity and another external contract can build a social network, forming strategic treaties between individuals.

Even such arrangements may not guarantee to provide any fruitful knowledge. It depends on the openness, trust, and knowledge sources of other companies. For example, certain Chinese companies may not be open and ready to disseminate knowledge by acting selfishly and not trusting even their own employees. In case a Chinese corporation shares its knowledge, it often treats subsidiaries as masters treat apprentices. A popular Chinese saying, "Farmers prevent their fertilizers from flowing into the fields owned by their neighbours" clearly demonstrates that [12].

Generally, there can be numerous barriers in transferring the very nature of knowledge, including economic reasons and individual behavioural patterns.

Russian companies' work environment can lead to the same consequences as the above-mentioned Chinese experience, albeit originating from different sources. Some of them may manifest in fear of losing social value as team leaders, for example. The atmosphere of not trusting anyone and a serious fear of loss of superior status leads to the behaviour of concealing information.

Continuous information sharing in local diverse teams of workers improves efficiency, giving real competitive advantage on the market, reduces redundancy of information, thus helping all members improve knowledge and strengthen teamwork. As a result, employees properly perform their jobs and follow their community responsibilities. Therefore, primary managerial tasks focus on providing the right infrastructure for knowledge transfer, building a trustful environment, aligning

S. V. Zykov, *IT Crisisology Casebook*, Smart Innovation, Systems and Technologies 300, https://doi.org/10.1007/978-981-19-2231-2_7

the right incentives. However, in case the managers are alienated from the team, the information flow stops and the employees desperately try obtaining the information required to complete their job tasks from random sources.

This section analyses factors and compares conditions in working environments that may help or hinder knowledge transfer and suggests improvement strategies.

The specificity of knowledge transfer process

Knowledge is a fluid mix of experience, values, contextual information, and expert insight that provides a framework for evaluating and incorporating new experiences and information. In 1977, the concept of knowledge transfer was coined by Teece [15]. The process of knowledge transfer can be understood differently. We are interested not only in transferring information between business units, but also in ensuring that it serves certain purpose and promotes business activities. As such, *knowledge transfer* is the process of transmitting information that has the source and destination actors and specific context, which serves the purpose of applying and using information to create competitive advantages for the company. The key objective of knowledge transfer is closely related to improving inner operational performance.

Knowledge can be tacit and explicit. However, we do not consider pure tacit or pure explicit knowledge as the real-world cases typically contain a mixture of both types.

There are numerous difficulties with the process of knowledge transfer. There is no evidence whether the process of learning happened or not; this fact particularly relates to tacit knowledge. Certain researchers attempt to measure that effectiveness through performance, innovation, and overall knowledge. However, it is nearly impossible to measure knowledge through performance as it is hard to monitor and manage factors related and unrelated to knowledge.

The difficulty of transferring knowledge is contingent on the tacitness of knowledge being transferred [7, 13, 16]. There is tacit knowledge that, being transferred from person to person, cannot be strictly standardized or specified, leading to problems with identifying the volume of knowledge and many other attributes. Therefore, relations between transfer source and recipient are important. Sometimes, there is a social gap of some kind between them; consequently, only a part of the initial amount of knowledge can be transferred.

Not all knowledge is easy or cheap to share. Considering technology or innovations, it is questionable whether transferring technology can be considered effective compared to developing that. In some cases, peer members of the workspace do not recognize the right incentive to share information. If knowledge is strategic and can benefit its holder sooner or later in the future, what can be the reason to share that, especially if methods or tools developed ultimately benefit a single person or business? Also, employees can face the fear of knowledge appropriation, which leads to trust issues in a workspace.

Another factor is the willingness and capability to absorb new information and learn. We address the learning-related factors in more detail in Chap. 5. Subsidiary power is affected by its absorptive capacity and provision of training. That has to be considered as one of the most influential factors to the absorption ability. So, for developing the absorption ability, there is a need for proper training and the process.

The process has not been shaped exclusively for new members only; a managerial and technical staff can also take advantage of it. Companies that practice regular or cyclic training, motivate employees to consume and transfer knowledge more effectively. In short, there is a dire need in systematic organization and novel approach to knowledge accumulation.

An important factor is consistency of the knowledge transfer process, as some companies do not value and appreciate knowledge sharing. Also, there are certain communication challenges when the companies fail to convey their knowledge sharing ideas effectively. This may originate from a plethora of problems, all of which inevitably make the transfer considerably painful and slow, and even trigger a crisis.

In spite of multiple challenges and problems, the world continues to collaborate with the help of international companies and cooperative multinational projects. As current experts retire carrying away their unique knowledge, it becomes essential to develop a uniform framework of knowledge transfer for different professional and international environments as shown in Fig. 7.1.

The framework suggested consists of two key parts: the knowledge transfer process, and a pyramid of factors helping or hindering its sub-processes. Among the transfer inhibitors are:

- Time
- Perception of information
- Importance of information
- Clarity of information
- Job role
- Training

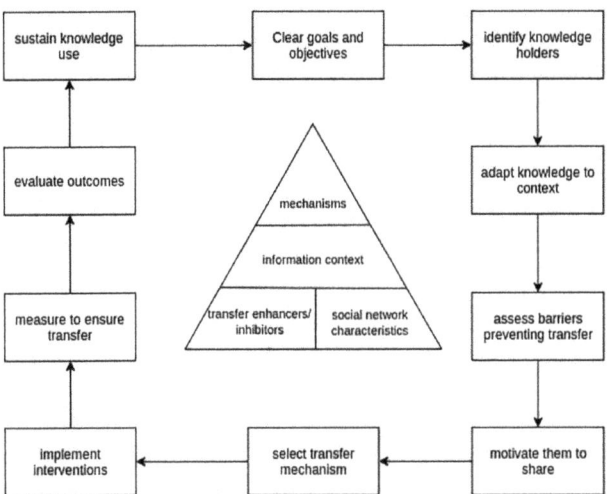

Fig. 7.1 Knowledge transfer framework

- Knowledge as a prerequisite for understanding.

Every enhancer should be considered separately for each individual case together with the following social network attributes:

- Structure
- Accessibility
- Openness
- Existing relationships between workers.

At this point, the social attributes matter as they promote high degree of openness, ability to threaten colleagues in different ways, and existing links between business units and individuals including cooperative projects and diverse companies. The organization structure, regulations, and traditions can say a lot about effectiveness as "knowledge transfer is completed in a certain context" [8].

Mechanisms of transfer knowledge include:

- Networking
- Face-to-face meetings
- Training off the job
- Training on the job
- Books, guides, and reports
- Best practices
- Databases and software
- Apprenticeship and mentorship
- Open spoken communication
- Written communication
- Colleague experience
- Co-location staff.

Overall, knowledge transfer mechanisms are divided into two categories: codification and personalization. Codification refers to translating information in documents, schemes, pictures, etc. by means of indirect, non-interactive techniques, which are adequate for information reuse, but less effective than personalization.

Further, the knowledge transfer framework provides a roadmap to identify differences in locally specific working environments, such as Russian and Chinese.

Working attributes around a Chinese environment

Chinese businesses rely on many ways of interpersonal relations. Factors of willingness and trust often play a more important role. Historically, competition between businesses was stiff, and employees cared about not being overperformed or pushed out of the market. Even within a corporation, departments may compete with each other making similar products as it happened at Microsoft during the "Ballmer's era" (see Chap. 5). It seems that the management wishes to promote a competitive spirit, always looking for ways to reduce costs and increase efforts. In such conditions, departments of huge enterprises work as one team, accumulating knowledge as it happens at Microsoft during the current "Nadella's epoch". Chinese working environment strives for collectivism, open spaces, little personal space, and over-performing. Having more personal contacts with a high degree of trustworthiness is considered more valuable than having more skills or talents. In this situation, striving to create more contacts and achieving collective values becomes paramount. Part of common practice is not to ask anyone for a favour, but to show proactivity, loyalty to the workplace, and leadership. Discussions on things already settled, an order of completion, or ways to express feelings rarely happen and are generally standardized by the national corporate culture.

Chinese employees rely on traditional Confucian values of "saving face" by using shades of grey rather than black and white as it currently happens at Huawei (see Chap. 5). Decisions are seldom made in a hurry and depend on many factors, although maintaining the company's strategic vision. Chinese employees have problems with completing tasks on time performing all steps with high quality; therefore, the leadership is often paternalistic.

Another basic principle is "Guanxi", which usually applies to people of the same status group having many actual contacts, and keeping relations, while having little or no direct interactions. This principle leads to regular transactions of favours. If someone did a favour, it supposes that the other side "owes" something in return. In the business sphere, some privileges of having such favours can be getting inside business information, trends, market insights, and the easing of performing operations. A formal approach is not common and is challenging to many workers.

One of the most important factors is trust and here, the Chinese rely on someone in the context of their families, education, and friends depending on the mutual image of the "face" that was created. In case persons cannot trust each other, they negotiate strategically. While the level of trust is high inside the group, they speak gently and carefully. Before someone can be trusted, it is important to assess how communication is going, as building just business relationships seems ineffective. Therefore, the Chinese strive to create or construct values that serve as solid ground for trust build-up. For instance, corporations can and declare themselves as poverty fighters and world savers, at the same time omitting sensitive details and even hiding potentially compromising data. Conversation results depend on many factors, and spending a lot of personal resources helping someone is often considered troublesome. Without strong communicative ties, a person asking for help can end up being gossiped about by the colleagues many times in a circle.

Heterogeneous cultural environments often require extra flexibility due to limited language proficiency and diversity of traditions and customs.

Working attributes around a Russian environment

Russian workstyle has been heavily influenced by the Soviet labour principles. Moreover, even new generations have been influenced by the same principles still embedded in the Russian labour ethics. Personal relations play a very important role. Slowly, Russians are becoming more individualistic; however, their collectivist culture often predominates. Strong division of personal and working activities intrinsic to state of mind, corporate holidays, and outs are seldom encountered. Historically, people learned to keep sensitive information to themselves because of the fear factor. Divulging negative or sensitive information can put them in the spotlight, create misinterpretations, and cause troubles, at least in the minds of the employees.

Usually, factors that can prevent a Russian from sharing knowledge are the difference in rank and fear of losing their status and competitive advantages. Generally, Russians need personal space and the opportunity to choose, whether to communicate or not. Accumulation of knowledge is an important competitive factor, considered by management and individuals. Consequently, they share little of it, particularly being uncertain of the possible benefits. People think to themselves "What do I gain from it?" In consulting positions that often include teaching or mentoring, Russians do not tend to spend much of their resources delegating green horns to documentation and other non-personal ways of conveying knowledge. Some employees think that only individual practice gives working knowledge and experience; they pay little attention to the knowledge diversity.

Employees barely see benefits of mentoring someone; instead, as the trainees leave the company all their efforts can go in vain or, in their opinion, they can spend that time more productively. Also, sharing vital knowledge can decrease individual value in the team. The difference in positions can be a reason to keep the information as they can avoid working with colleagues in lower positions or sharing knowledge. In case of a critical problem, Russians often rely on their relationships. In many cases, they often prefer to either convey these problems to the supervisor or consult with the boss. At the same time, discussing problems in Russian society is a taboo. As such, in certain cases employees try to solve these challenges by hiding, redirecting to other issues, or offering inadequate solutions. Such behaviour patterns lead to a lack of feedback, distorted perception of reality, and often conceal problems, making it hard to develop the right solutions.

Trust is a basic requirement for sharing any important knowledge. Before accepting newbies, Russians tend to test them on simple tasks or spend much time to build a trustful connection. An employee should demonstrate personal loyalty and understanding in the key principles and managerial activities.

Comparison and challenges

According to research by Li, Chang et al., performed among 23 countries, Russians are more individualistic and have greater power distance compared to the Chinese [7].

Figure 7.2 shows that Russian supervisors usually command rather than lead the team. Fear of sharing negative activities is greater than in the Chinese case, and knowledge transition from supervisors seems less effective compared to Chinese due to greater power distance. Desire to transfer knowledge decreases when power distance grows; this harmfully affects digitalization, initiation, absorptive capacity, and network transfer density. Higher individualism does not promote knowledge transfer: individuals tend to be more independent, relying on their own experience and focusing on their personal goals. This is the opposite to Chinese collective interest and responsibility driving them to active knowledge sharing.

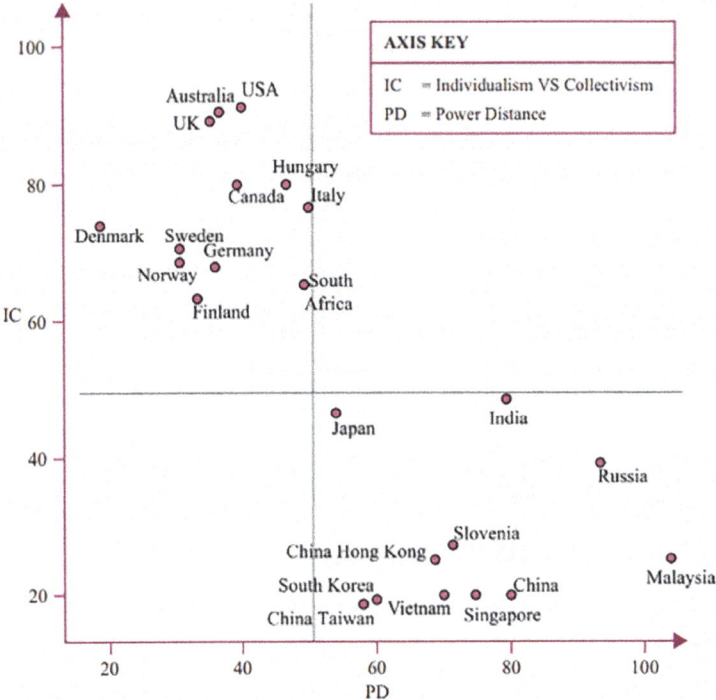

Fig. 7.2 Individualism versus collectivism; power distance

Both Russian and Chinese cases have clear specific features. Let us compare them and identify strong and weak points using the knowledge transfer aspects of the IT Crisisology framework.

First, clarifying goals and objectives, Chinese tend to be more accurate, they focus on completing tasks in time with good enough quality and limited resources. They are used to setting clear and simple goals and tasks dedicated to each team member.

As a rule, Russians often experience problems with quality feedback while solving challenging situations.

Identifying knowledge holders is similar for both cases as they tend to obtain such information through communication with the superiors.

Adapting knowledge to context can take more time for Russians, since tasks are not always clearly defined and may overlap with other competencies. Chinese also used to work with a high degree of uncertainty in the past; however, the context of their current tasks is narrower.

Motivation to share knowledge can be a tougher problem for Russians, especially in case the information required is beyond their duties. Chinese can be motivated more easily, due to high competitiveness, cultural, and traditional specifics.

Measuring knowledge transfer is rare for Russian or Chinese traditions. Usually, the knowledge recipients display their interest in such activity. In case of receiving wrong information, Russians usually try to understand the issue and dig deeper on their own before asking questions. Chinese often behave similarly.

Conclusion

In this section, we compared the factors influencing knowledge transfer in Russian and Chinese working environments. Our analysis exposed strengths and weaknesses of the process. Research demonstrated that trust and personal contacts are helpful in both contexts, whereas individualism and power distance are among the inhibiting factors. Strong team loyalty results in enclosed fractions in the company, where the members are unwilling to share knowledge with the outsiders. Globalization processes require intensive knowledge sharing and managers follow this trend, compromising and harnessing diverse cultural factors.

7.2 Taming the Tiger: Soft Skills

More than coding: five soft skills a developer need

The best programmers today do not just write code. What "flexible skills" do developers need and how to master them?

Developers' career largely depends on the ability to work in a team, build relationships with colleagues, and quickly and correctly understand customers. It is necessary not only to develop a product, but also to be able to present it and defend one's decisions. To be successful and climb the career ladder, IT pros need to focus on developing the following skills.

1. **Empathy**

The ability to understand other people is often underestimated. But it is empathy and emotional intelligence that play a crucial role in teamwork. Any developer should be able to put oneself in the shoes of colleagues—this will make it easier to exchange ideas and not oppress colleagues when something goes wrong. A high level of empathy also helps to better understand customers. For example, a developer might think that the code produced is working fine and that is enough, but ultimately one should always think about the convenience of each potential user of this product.

How to learn/what to remember

First, one needs to listen carefully to other people. Stay friendly and open. Let the coworkers know they can always ask for help or advice. Trusting relationships are the basis of a healthy atmosphere, and only such teams have every chance of successful development.

Soft Skills Engineering is a relevant podcast on the topic [14]. One can also listen to certain episodes of the Developer Tea podcast and read the books: "Practical empathy: for Collaboration and Creativity in Your Work" by Indy Young and "Mirroring People: The Science of Empathy and How We Connect with Others" by Marco Iacoboni [3, 5, 18].

2. **Stress resistance**

Resilience to stress is a skill that can save a lot of nerves. Developer's work is stressful: the code does not work for no apparent reason, colleagues do not understand the requests, and customers demand the impossible. It is important to remain calm in any situation. Skilled developers know how to give feedback not getting personal, although repeating the same comments several times a day.

How to learn/what to remember

For a team lead, the main goal is creating an atmosphere where all team members can honestly express their opinions. Weekly retrospective sessions work very well, where everyone gives feedback on the work done: what turned out well, what difficulties were faced, and where help is needed.

A certain amount of patience is required when explaining technical details to colleagues or clients who know little about programming. Allocate sufficient time for these meetings. It is better to spend a few hours to explain everything in detail, than to be annoyed later that someone does not understand obvious things.

Practice in feedback: address not only mistakes, as often happens, but also positive aspects.

Identify a few strengths, focus on their development, and keep the rest at an acceptable professional level.

3. **Teamwork**

Working in a team involves key "soft" skills for any developer. There is a team of experts behind any smart IT product, and the speed of development directly depends on the ability of all participants to work harmoniously. Developers must know their colleague responsibilities to get advice. It is important to follow the rules and not get carried away with self-expression: an initiative is useful only if and when it works for common goals.

How to learn/what to remember

Remember the rules and follow them clearly.

Do not be afraid to take responsibility for specific tasks and make decisions in the domain of expertise.

Developers often collaborate in diverse teams, including multinational and geographically distributed ones. It is therefore important to express the ideas clearly and have an acceptable English level. To use professional vocabulary with confidence, study the technical terms and take an English course for IT professionals. YouTube Videos are helpful to practice speaking; one current example is Marina Mogilko's blog [9].

4.   **Negotiation**

It would seem that negotiation skills are unlikely useful to those who spend hours in coding without talking to people. However, developers of the state-of-the-art companies communicate intensively with their colleagues, managers, and customers. For instance, a developer can be brought in as an expert for a meeting with a customer. Negotiations also help in day-to-day work: when choosing a technology, prioritizing the backlog, and evaluating results. Developers should be able to defend their viewpoints.

How to learn/what to remember

Take the online free course, Successful Negotiation: The Strategies and Skills Needed from the University of Michigan [11].

Developers can master the negotiation skills on their own as many books have been written on this topic. Among the most popular and useful are "Kennedy-on-Negotiation" and "Negotiating Without Defeat" by Fisher, Urey, and Patton [4, 6]. To practice, be proactive in meetings, participate in meetups and conferences, and your confidence and skills will grow with each new performance.

How to learn/what to remember

Developers often have to switch between different tasks as it is easy to get procrastinated or burn out in the workplace. In this intensive event flow, it is important to manage time properly. Purposeful time management increases work efficiency without getting exhausted.

5.   **Time Management**

Plan the day in the morning or the night before. Do not underestimate the ordinary to-do list to monitor activities. Start any day with a "frog", i.e., activity that must be done although one would prefer deferring it. Such tasks require a substantial amount of energy, even though sent to "background": they hang over, hindering concurrent activities efficiency. In fact, these "frogs" are not as unpleasant as imagined. Due to procrastination, such tasks are often treated negatively; therefore, it is recommended to manage them early in the morning.

The Pomodoro Technique

The method is simple: after a 25-min period of nonstop work, take a five-minute break. Further, after four such work cycles, rest for another 20 min [2]. To track time, use a timer app such as Move On Productivity timer available at the AppStore [10]. There are many books written on the topic of time management, among the most popular is Allen's "Getting Things Done, GTD", and "Maximum Concentration [1]. How to maintain efficiency in the era of clip thinking" by Wooten [17]. In the new digital era, the growing demand in "soft" skills proves that "human" qualities become vitally important along with professional ones. Such qualities as teamwork and add-value motivation may become a critical competitive advantage over the developers with the same set of the core "hard", i.e., professional skills.

7.3 Conclusion

This Chapter elucidated the subtle "human factors" that may either promote or prevent the digital transformation processes. These factors are mission-critical, being an essential "pillar" of the ITC framework. They typically result from client-to-developer miscommunications and manifest themselves as crisis triggers. The management attitude to these, however, often centres on neglect or even ignorance. First, we studied the process of knowledge transfer between the digital product developers and their clients to systematically address mission-critical factors of digitalization and sustainable development. Our communication model included the informing science approach based on Shannon information theory, and the "seven principles" of knowledge transfer, being essential ingredients of the ITC framework. Further, we outlined the development practices for the "soft" skills. These are human abilities that harness tricky human-related factors, promote digitalization, help preventing crises, and assist in sustainable company development.

References

1. Allen D. (2015). Getting Things Done, GTD. Retrieved February 17, 2022, from https://gettingthingsdone.com/.
2. AttendanceBot Blog. (2020, August 20). Retrieved February 17, 2022, from https://www.attendancebot.com/blog/pomodoro-technique/.
3. Developer Team podcast. (2022). Retrieved February 17, 2022, from https://developertea.com.
4. Fisher, R., Urey, W., Patton, B. (n.d). *Negotiating without defeat*. Retrieved February 17, 2022, from https://sato-1.ru/en/repairman/rodzher-fisher-uilyam-yuri-bryus-patton-peregovory-bez-porazheniya/.
5. Iacoboni M. (2009). *Mirroring People: The Science of Empathy and How We Connect with Others*. NY: Picador. ISBN-10 0312428383.
6. Kennedy G. (2016, November 15). *Kennedy-on-Negotiation*. Routledge. ISBN 9781138263147.
7. Kogut, B., & Zander, W. (1992). Knowledge of the firm, combination possibilities and technology replication. *Organization Science, 3*, 383–397.
8. Li, J. H., Chang, X. R., Lin, L., & Ma, L. Y. (2014). Meta-analytic comparison on the influencing factors of knowledge transfer in different cultural contexts. *Journal of Knowledge Management, 18*(2), 278–306.
9. Marina Mogilko's blog. (2022). Retrieved February 17, 2022, from https://marinamogilko.co/.
10. Move On Productivity. (2022). Retrieved February 17, 2022, from https://apps.apple.com/us/app/move-on-productivity-timer/id1323419635.
11. Online free course. (2022). *Successful Negotiation. The Strategies and Skills Needed*. University of Michigan Retrieved February 17, 2022, from https://online.umich.edu/courses/successful-negotiation-essential-strategies-and-skills/.
12. Ramasamy, B., Goh, K. W., & Yeung, M. C. H. (2006). Is Guanxi (relationship) a bridge to knowledge transfer? *Journal of Business Research, 59*(1), 130–139, ISSN 0148-2963.
13. Simonin, B. L. (1999). Ambiguity and knowledge transfer in strategic alliances. *Strategic Management Journal*.
14. Soft Skills Engineering. (n.d). Retrieved February 17, 2022, from https://softskills.audio.
15. Teece, D. J. (1977). Technology transfer by multi-national firms: the resource cost of transferring technological know-how. *The Economic Journal, 87*, 242–261.

16. Wang, P., & Singh, K. (2001). *Determinants and outcomes of knowledge transfer: A study of MNCs in China.*

17. Wooten. (2022). *Maximum Concentration. How to maintain efficiency in the era of clip thinking.* Retrieved February 17, 2022, from https://www.amazon.com/Maximum-concentra tion-maintain-efficiency-thinking-ebook/dp/B09JCCY8RP/ref=sr_1_7?qid=1639495170& refinements=p_27%3AJames+Wooten&s=books&sr=1-7.

18. Young, I. (2015). *Practical Empathy: for Collaboration and Creativity in Your Work* (1st ed., 259 pp.). Publisher: Rosenfeld Media. ISBN-13: 978-1933820484.

Conclusion: Post-digitalization: What Next?

In our study, we investigated case method application to IT-intensive business development. This method, as a part of the IT Crisisology framework, proved a powerful tool not only for training but also for research, particularly in an uncertain environment, i.e., in a crisis.

Let us use a metaphor to convey the lessons learned while analysing the case studies and discussing them. One of the metaphors for the business ventures we used in our previous books was related to dinosaurs and other extinct creatures. The idea was that any business evolves in a certain environment and deals with competitors who are often aggressive. However, the changing environment itself is often a threatening factor to business development and a major challenge. Any business is likely to expire unless it quickly adapts to these environmental changes. We compared the large-scale, enterprise-type businesses with dinosaurs as these beasts were huge and complex creatures slowly responding to critical environmental changes, i.e., a crisis.

Another interpretation of such an environmental condition shift is related to the changing dinosaur's perception (i.e., its ability to become aware of something through the senses), which is related to the knowledge transfer processes that we discuss in Chap. 7. This knowledge transfer relates the ancient predator to the environment in terms of how the business operates (i.e., understands, analyses, interprets, etc.) the key ideas of the product to be delivered to the market to meet the customer's requirements and expectations. Successful transfer of these initially vague ideas into certain product features that match customer's expectations requires developing a set of new abilities, i.e., soft skills to adapt to the new and continuously changing client's behaviour. This process of knowledge transfer and product development is somewhat similar to a dinosaur hunting for food, which depends on many factors such as daytime, weather, surroundings, etc.

As the Introduction of this book suggests, to convey a meaning of a certain complex and/or dynamic object, we often refer to such genres as stories, parables, myths, legends, fables, fairy tales, etc. Let us recall a famous Russian satirist and journalist,

Fig. 1 The quartet

Ivan Krylov (1769–1844), the author of nearly 250 fables, most of which allegorized certain types of people, and their traits of character, as animals. One of Krylov's fables, The Quartet literally deals with a problem of orchestration of a complex system (see Fig. 1).

The story starts as follows (translated by Olga Dumer) [1]:

One summer day, a monkey,

A goat, a bear, and

A donkey

Got the idea to perform as a quartet.

They got a cello, violins, and music scores

And settled comfortably outdoors —

To charm the audience with gentle minuets.

They struck the chords with ample force, and yet

No tune, just raspy squeaks.

Having read this book, it is evident that the reason for this unsuccessful orchestration is the diversity, and certain "soft" skills are required to overcome this crisis. One of such mission-critical skills is teamwork, which we address in Chap. 7. Without proper training and applying the "seven principles", including Practice and Feedback, it is impossible to master such a complex competence as playing an orchestra. Even a surface-level environmental change, although being important, is still clearly insufficient to reach this ultimate goal of orchestration:

"Your ears are not keen for music scale",

Replied the nightingale with a sweet warble,

"The skills are key to music, not your squabble.

Your changing seats, or places, or positions

Will not help you, my friends, become musicians".

Let us peruse through more examples of animal-based allegorizing to understand the outcomes of the business development in crises. The plot is based on the case of Huawei, an extremely successful Chinese telecom product maker. Surprisingly, Huawei survived in a number of crises resulting from the lack of capital, technology, and HR. Therefore, their crisis aspects included all the three ingredients of our ITC framework, such as business, technology, and human-related factors. Nevertheless, they dared to challenge and successfully compete with the major Western manufacturers, nearly 90% of which had been established for over a century. The Huawei Story says: "They were playing a game of ants versus elephants, and only crazy dreamers could believe that they would eventually see victory" [2]. Being much younger and smaller than their Western elephant-like neighbours, the Chinese "ants" competed with the "elephants". Due to the law of the jungle, less than 1% of these "ants" survived in this fight. Only four Chinese companies gradually evolved into "calves" and later became comparable to the Western "elephants", Huawei being one of them.

The story further describes an ant colony carrying construction materials for their nests to get prepared for the environmental disasters, i.e., crises. This resembles small and medium enterprises, which are important for any national economy, and particularly for Chinese as they make up some 60% of national GDP, around 80% of jobs, and nearly 50% of the domestic tax revenue.

In 2010, *CEOCIO Journal* in "Evolution from Wolves into Lions" suggested subdividing the telecom businesses into three kinds [2]. The first kind was "lions", i.e., world industry multinationals leading in technologies, products, capital, and management. These "lions" were typically located in the West; they were self-confident and looked superior against the backdrop of global competition. The second kind referred to "leopards", i.e., mid-size joint ventures. The third kind was "wolves", or local companies with relatively outdated technologies and lower quality products, although market aggressive and often customer focused. Huawei, historically being a typical "wolf", survived through the environmental selection and became a threat even to the "lions" of the West.

To identify our findings, let us revisit the structure of the book.

In Chap. 2, we introduced the digitalization phenomenon and approached it from the perspective of the IT Crisisology framework. We systematized the key definitions, ideas, and principles of company digital transformations. Also, we introduced case studies as the primary research method. Specifically, we applied the case-based reasoning for informed decision-making in uncertain, i.e., crisis-like conditions. Further, we applied this multi-faceted and practically focused approach to examine such a complex phenomenon as digital transformation and validate the ITC framework.

In Chap. 3, we analysed a few cases of large-scale and/or IT-intensive business transformations that happened in the pre-digital era. We identified environmental complexity as a potential crisis trigger and, therefore, recommended applying the ITC framework. Particularly, we investigated several mission-critical aspects, such as company structure, geographical range, and business diversity. The real-world case studies included Accenture, Cirque de Soleil, Dropbox, Disney, Foursquare, and Zara. These businesses operate in very different areas; however, the case method applied as part of the ITC framework served as a concept proof and reinforced the principles and practices introduced earlier as it revealed the strong and weak sides of the transformations and suggested helpful strategies and techniques.

In Chap. 4, we moved the timeline to the digital era and discussed transforming dynamic and competitive environments. The cases included publishing (IGI Global and Springer) and fast-food (Drinkit and Dodo Pizza) businesses. We concluded that changeable landscapes are likely triggers for crises, and carefully balanced digitalization in terms of business, technical, and human-related factors assists in successful crisis management and results in sustainable organizational development.

In Chap. 5, we investigated the digital transformation cases in such diverse and large-scale multinationals as Microsoft and Huawei. First, we discussed the multiple aspects of diversity, which typically impact business performance. For these IT-intensive and organizationally complex MNCs, we recommended the key drivers for transformation strategies, which critically depend on informed decision-making, smart and innovative technologies, diversity management, resonant knowledge transfer, and agile customer relationships.

In Chap. 6, we discussed the Russian Forest industry digitalization. The key problem for this important, large-scale, and very traditional industry that we detected was the process rigidity in the current very dynamic environment with rapidly changing technologies and regulations. Specifically, potential crisis factors originated from multi-dimensional complexity that included diversity challenges as well as organizational, legislative, and logistical problems. To manage this crisis essentially hampering digital transformation of the industry, we recommended tradeoff-based optimization strategy aimed at innovative development and agility improvement.

In Chap. 7, we studied the "human factors" that may help or hinder digitalization. These factors, which are one of the "pillars" of our ITC framework, typically originate from miscommunications in digital production processes between the client and the developer. Being subtle, these factors often activate crises; however top management tends to repeatedly neglect or even ignore them. To address these mission-critical factors for sustainable development and digitalization, we identified a set of "soft" skills, such as teamwork, negotiation, and time management, that drive digitalization and sustainable development and eventually make them succeed.

Let us summarize the research outcomes of the book.

Currently, digital transformation becomes mission-critical for smart and innovative businesses. According to our ITC approach, this transformation may result in a crisis of digital product delivery, i.e., being late, over budget, lacking essential functionality, or even undelivered at all. To manage this crisis, multiple criteria tradeoff-based optimization for business processes and product development are required. This optimization strategy should address the three "pillars" of the ITC framework, including business, technology, and human-related factors in a balanced way.

The essential ingredients of the approach recommended include the following:

- Lifecycle model including thorough analysis, continuous benchmarking, and adjustable planning as in PDCA/DMAIC
- Product development methodology including agile-based approaches

- Informed decision-making and communication including a two-way feedback loop with resonant compensation and "seven principles" of knowledge transfer
- Addressing gender, national, religious, and cultural diversity by means of harnessing "human factors" and improving "soft" skills
- Maturity management as prescribed by CMM/CMMI models
- Implementing success patterns and best practices and avoiding dangerous anti-patterns as illustrated by the set of business transformation case studies.

The above ingredients allow for complexity management, which assists in crisis-resistant development.

To provide food for thought, we adopted a case study-based approach, which included real-world stories of transformations in the leading companies ranging from relatively small startups to world famous multinationals. To make a comprehensive and statistically valid selection, we studied over a dozen cases of business ventures different in size, industry, location, and a few other aspects. Majority of these case studies are success stories of tough survival in crises. A few instances, however, are failure examples of pitfalls to avoid.

Based on the case method with no single correct answer due to multiple tradeoffs, we consider that it is up to the reader to draw conclusions regarding the applications to particular situations of the current and future digitalization trends and strategies embodied in the above stories.

We wish you every success in sustainable development and rewarding digitalization!

References

1. Krylov I. (2022). Quartet (O. Dumer, Trans.). Retrieved February 17, 2022, from https://ruverses.com/ivan-krylov/the-quartet/4137/.
2. Tao, T., & Chunbo, W. (2015). The Huawei Story. SAGE Publications.

Appendix A
The Case Method and Russian Informed Digitalization

See Figs. A.1, A.2, A.3, A.4, A.5, A.6, A.7, A.8, A.9, A.10, A.11, A.12, A.13, A.14, A.15, A.16, A.17, A.18, A.19, A.20, A.21, A.22, A.23, A.24, A.25, A.26, A.27, A.28, A.29, A.30, A.31, A.32, A.33, A.34, A.35 and A.36.

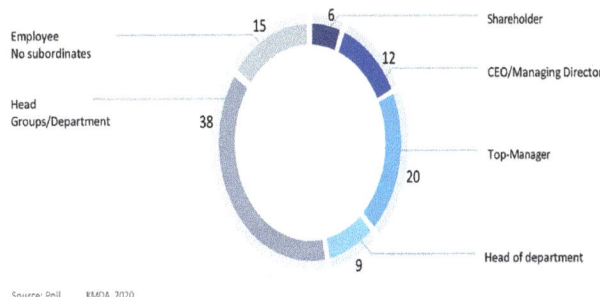

Fig. A.1 Respondent positions in their companies, % (*Source* KMDA, 2020)

Fig. A.2 Industry profiles of the respondents (*Source* KMDA, 2020)

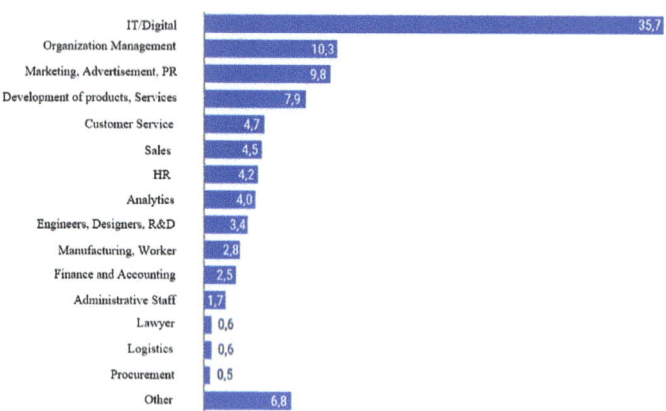

Fig. A.3 Professional activities of respondents in their companies (*Source* KMDA, 2020)

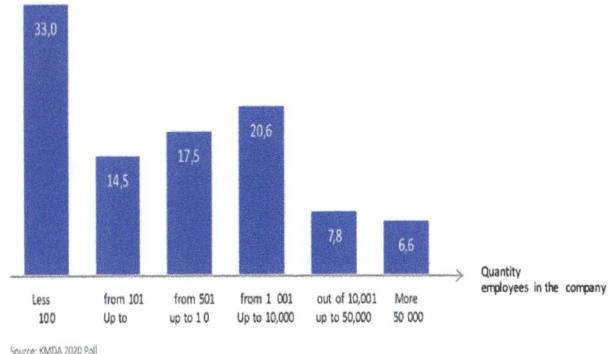

Fig. A.4 Respondents by staff number, % (*Source* KMDA, 2020)

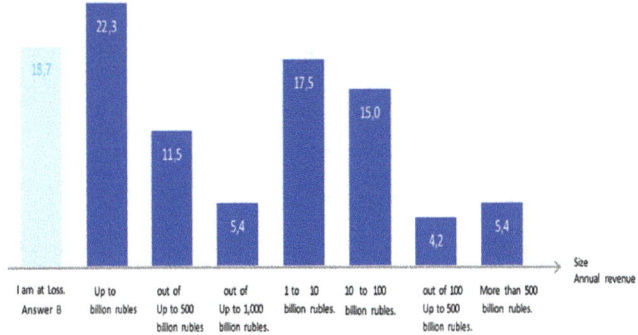

Fig. A.5 Total annual revenue of the respondent companies, % (*Source* KMDA, 2020)

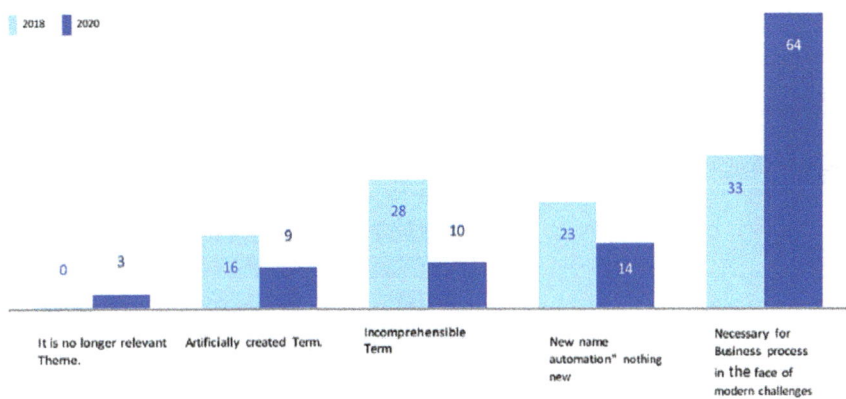

Fig. A.6 Personal perceptions of the digital transformation (*Source* KMDA, 2020)

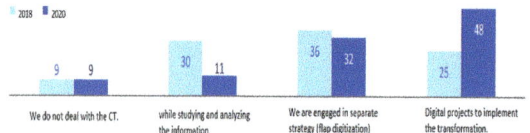

Fig. A.7 Digital transformation status in Russian companies, % (*Source* KMDA, 2020)

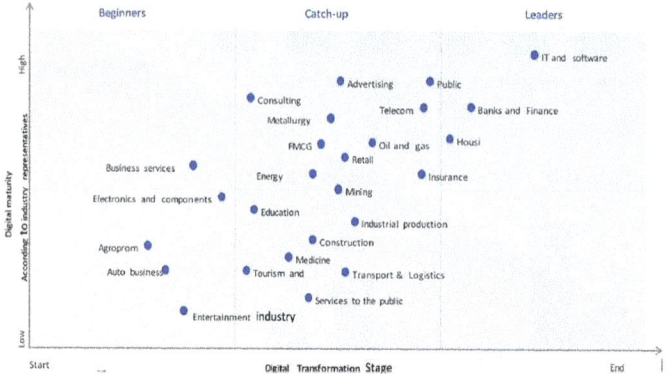

Fig. A.8 Digital transformation status by industry (*Source* KMDA, 2020)

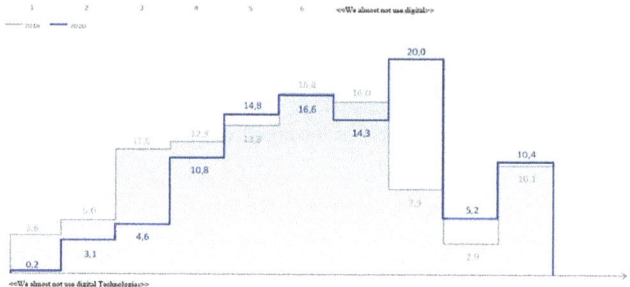

Fig. A.9 Company self-assessment of their digital maturity, % (*Source* KMDA, 2020)

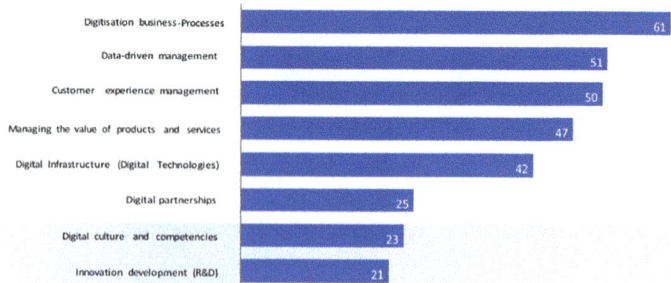

Fig. A.10 Digital transformation priorities of Russian companies, % (*Source* KMDA, 2020)

Fig. A.11 Digital transformation maturity priorities of Russian companies (*Source* KMDA, 2020)

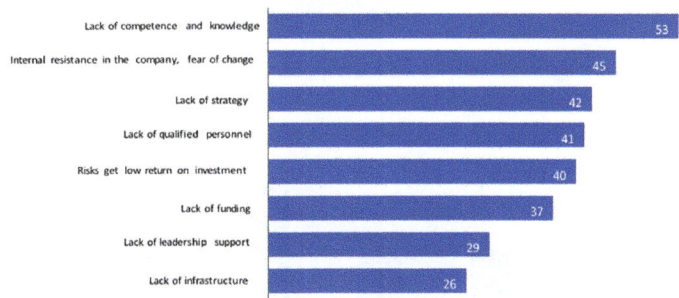

Fig. A.12 Key obstacles to digital transformation, % (*Source* KMDA, 2020)

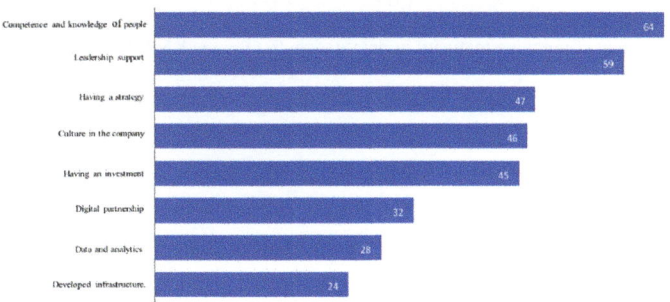

Fig. A.13 Factors of success of digital transformation (according to respondents), % (*Source* KMDA, 2020)

Fig. A.14 Project management methods (*Source* KMDA, 2020)

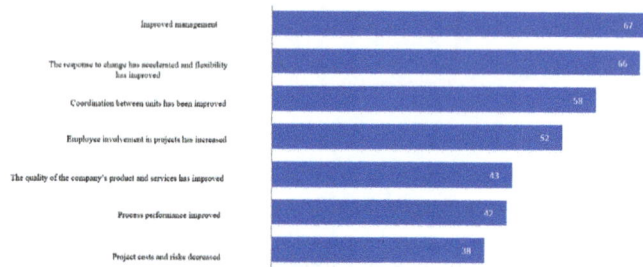

Fig. A.15 Project management methods used in respondents' companies, % (*Source* KMDA, 2020)

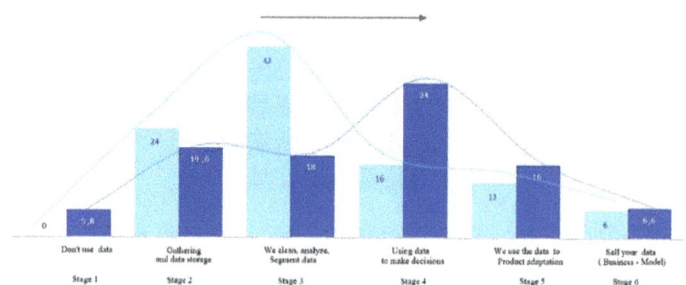

Fig. A.16 Russian companies maturity in data usage, % (*Source* KMDA, 2020)

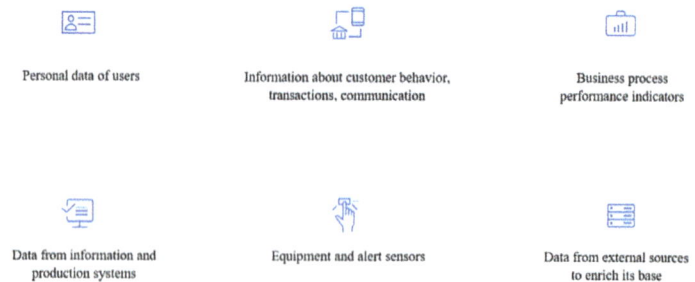

Fig. A.17 Main types of data used by Russian companies (*Source* KMDA, 2020)

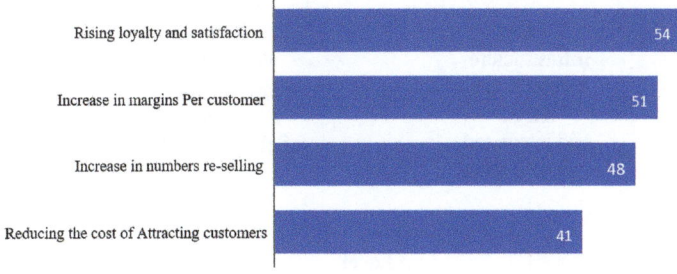

Fig. A.18 Key effects of customer experience management, % (*Source* KMDA, 2020)

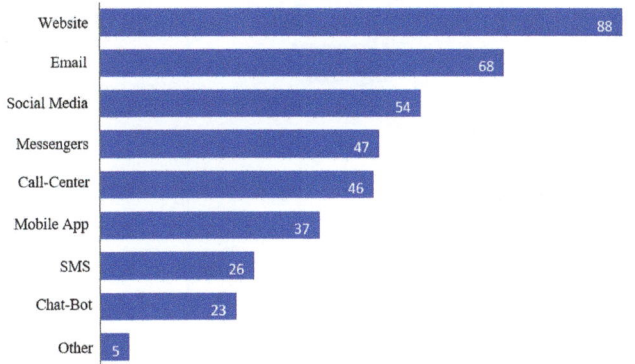

Fig. A.19 Digital channels of Russian companies, % (*Source* KMDA, 2020)

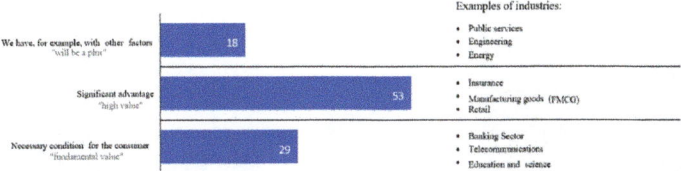

Fig. A.20 Values of digital services for customers of Russian companies, % (*Source* KMDA, 2020)

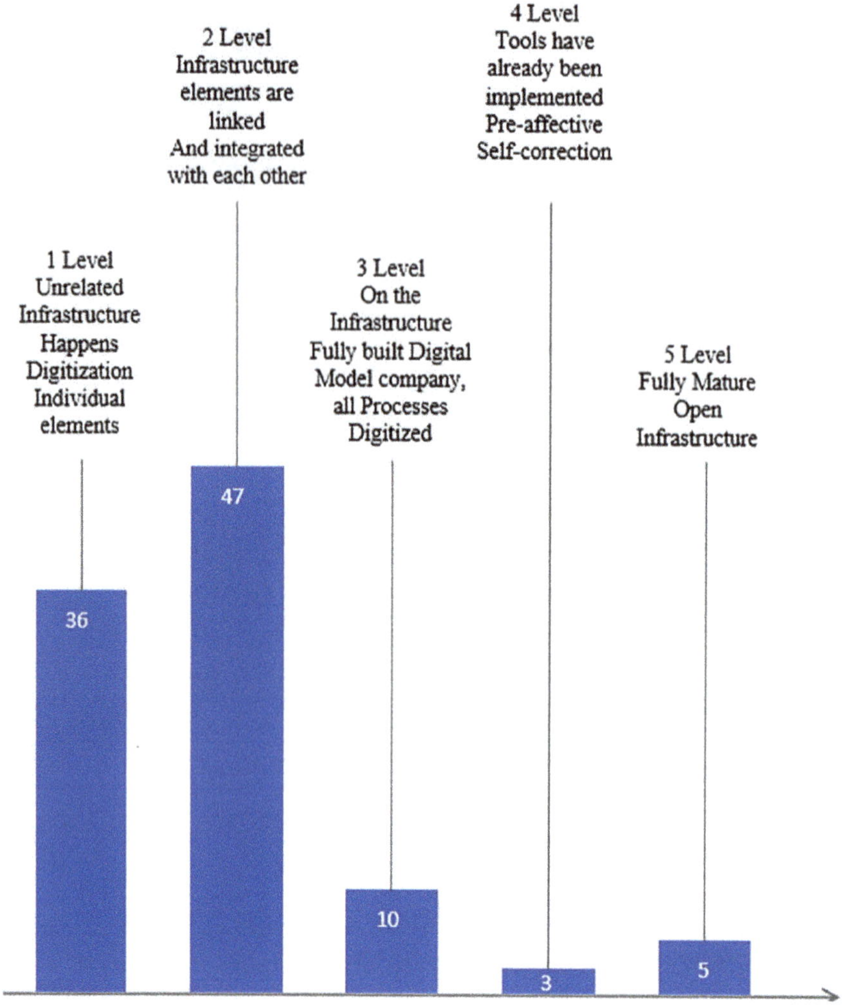

2 Level
Infrastructure
elements are
linked
And integrated
with each other
4 Level
Tools have
already been
implemented
Pre-affective
Self-correction
1 Level
Unrelated
Infrastructure
Happens
Digitization
Individual
elements
3 Level
On the
Infrastructure
Fully built Digital
Model company,
all Processes
Digitized
5 Level
Fully Mature
Open
Infrastructure
47
36
10
3
5

Fig. A.21 Digital infrastructure development levels in Russian companies, % (*Source* KMDA, 2020)

Possible Infrastructure ahead of business challenges
16
Infrastructure and business challenges evolve in sync
21
Infrastructure lags behind from business
55
It's hard to answer
8

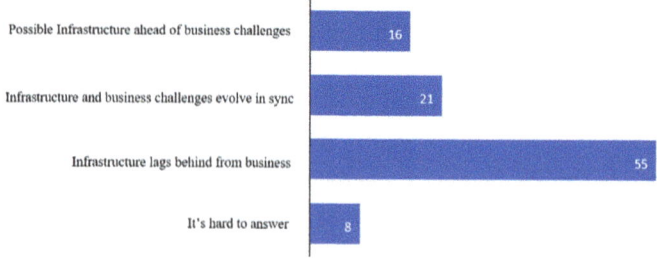

Fig. A.22 Matching digital infrastructure benefits to business objectives, % (*Source* KMDA, 2020)

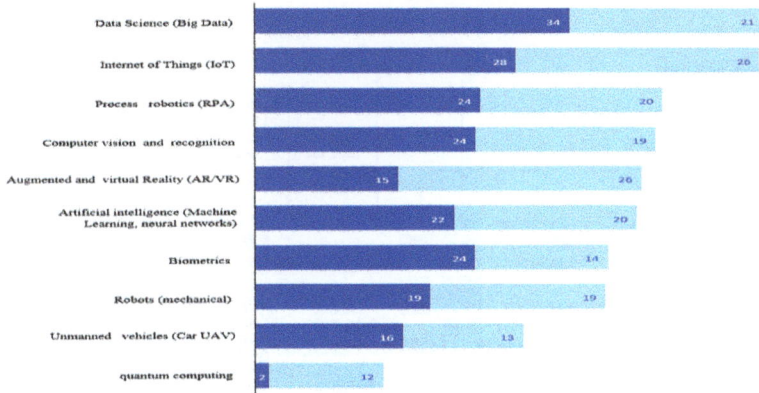

Fig. A.23 Technological trends implemented and planned by Russian companies, % (*Source* KMDA, 2020)

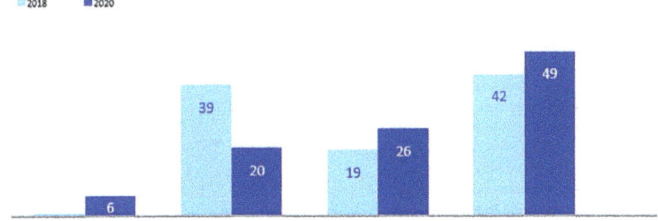

Fig. A.24 Products and services implemented through digital partnership, % (*Source* KMDA, 2020)

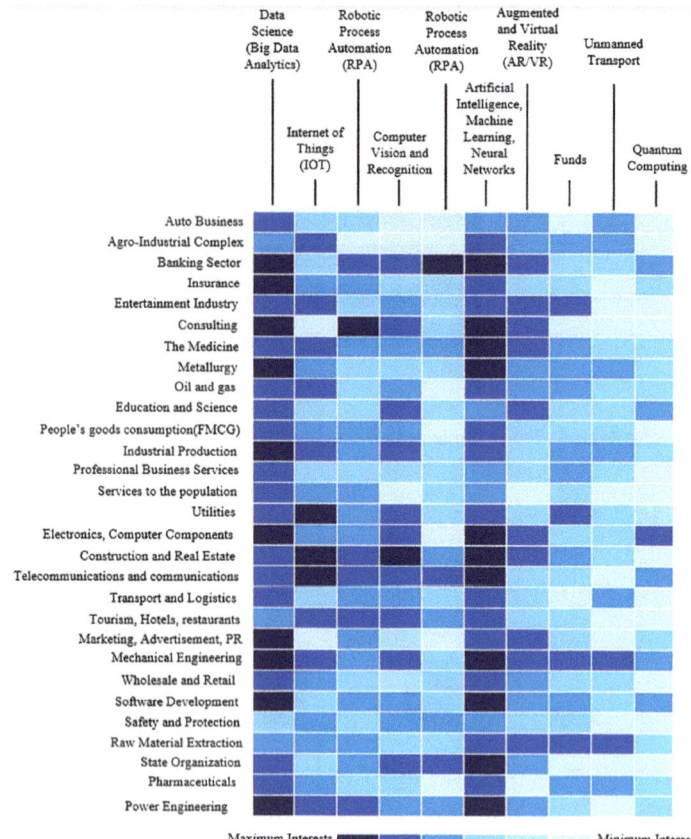

Fig. A.25 Hot areas for Russian companies in digital technologies (*Source* KMDA, 2020)

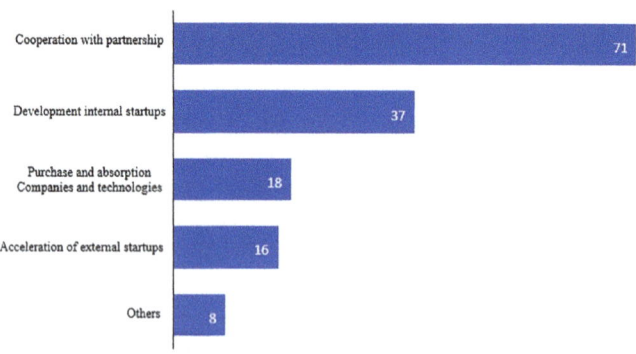

Fig. A.26 Company innovative practices, % (*Source* KMDA, 2020)

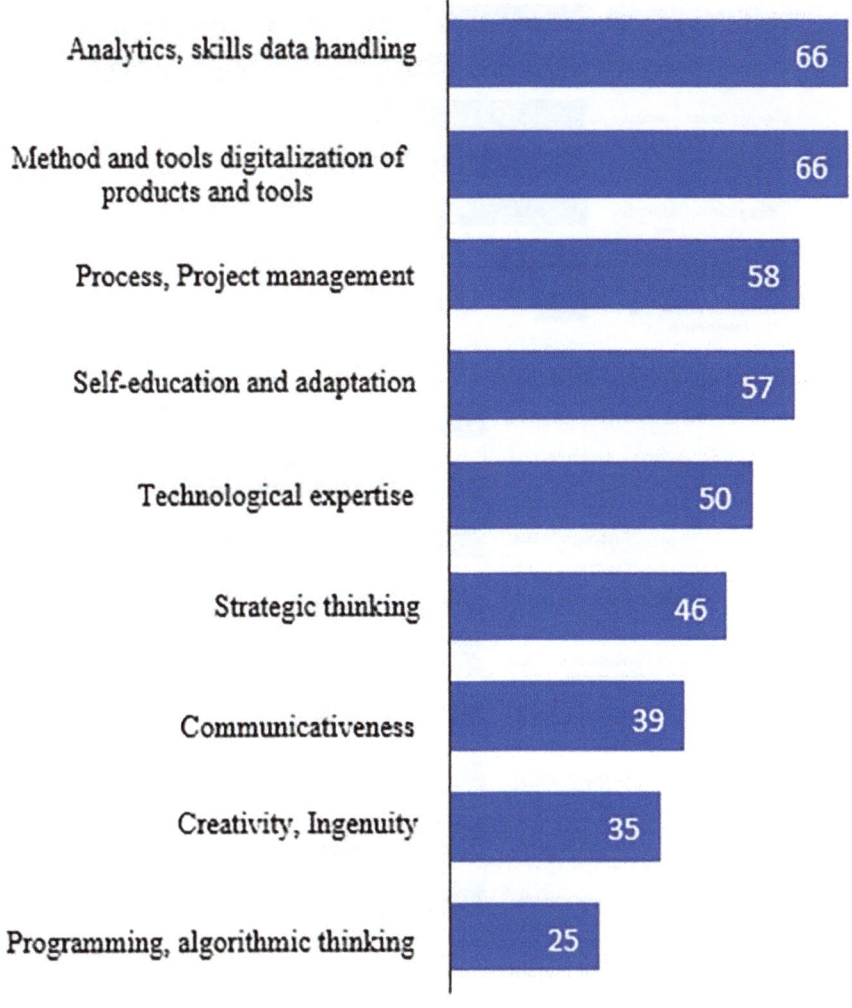

Fig. A.27 Key skills and competencies for digital transformation, % (*Source* KMDA, 2020)

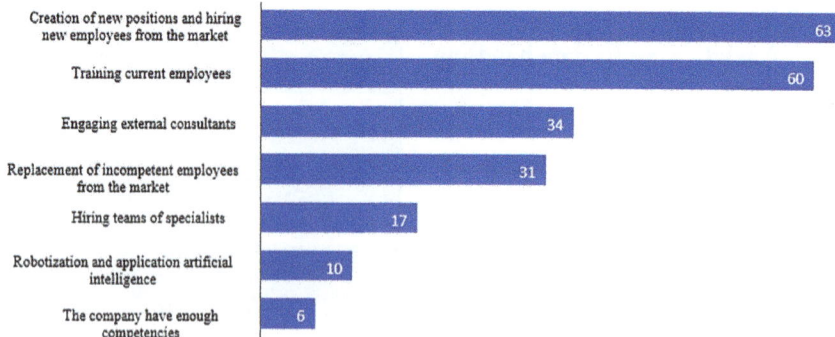

Fig. A.28 Dealing with insufficient skills and competencies, % (*Source* KMDA, 2020)

Fig. A.29 Prospective areas of employee training in digital transformation, % (*Source* KMDA, 2020)

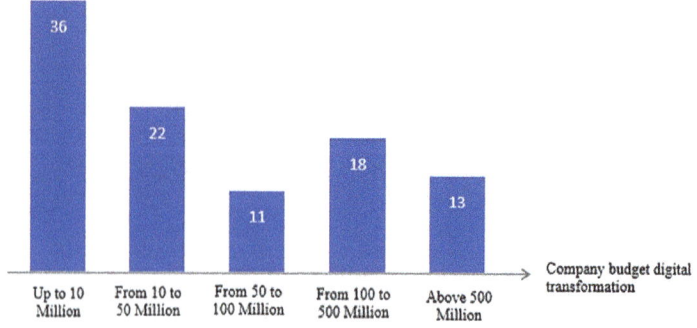

Fig. A.30 Budget of Russian companies for digital transformation, % (*Source* KMDA, 2020)

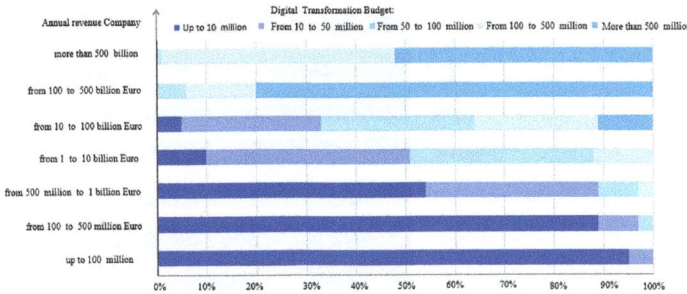

Fig. A.31 Digital transformation budget depending on the annual revenues, % (*Source* KMDA, 2020)

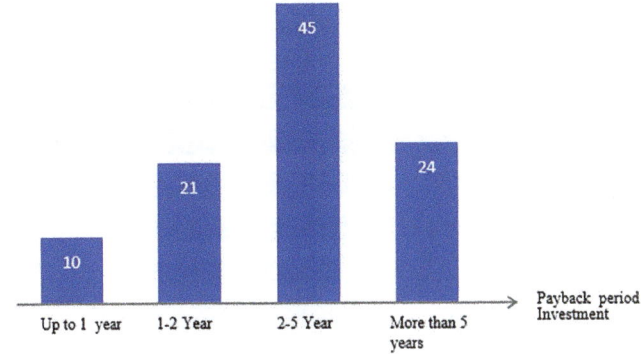

Fig. A.32 Expected return on investment in digital transformation, % (*Source* KMDA, 2020)

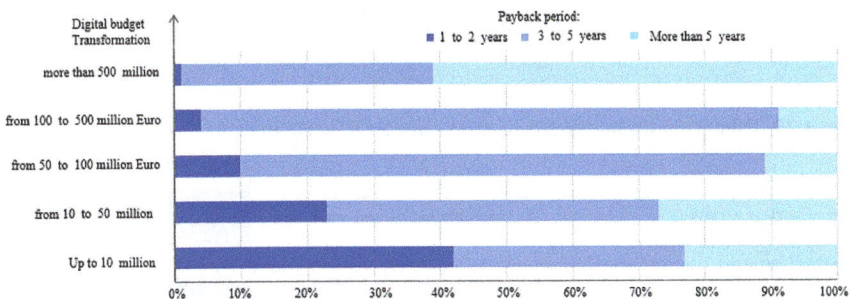

Fig. A.33 Expected return on investment for digital transformation depending on its budget, % (*Source* KMDA, 2020)

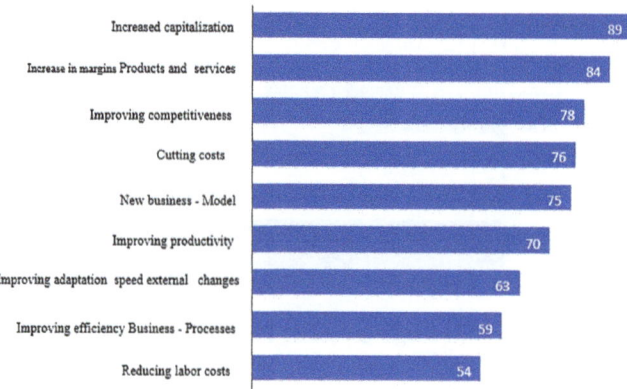

Fig. A.34 Russian CEO expectations from digital transformation, % (*Source* KMDA, 2020)

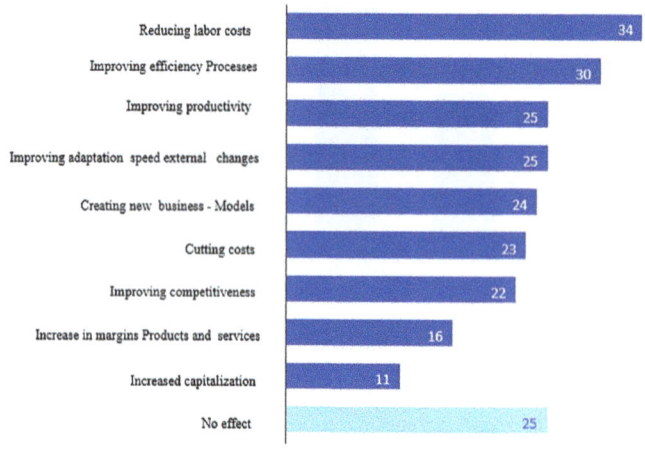

Fig. A.35 Digital transformation effects, % (*Source* KMDA, 2020)

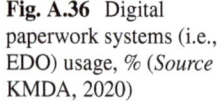

Fig. A.36 Digital paperwork systems (i.e., EDO) usage, % (*Source* KMDA, 2020)

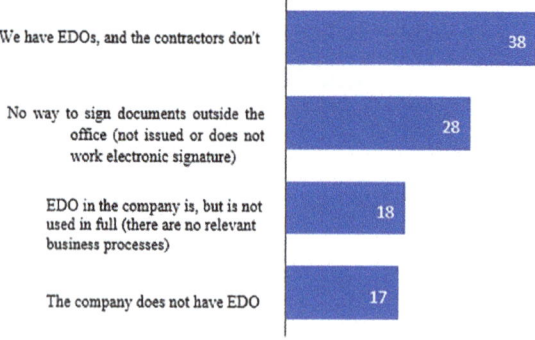

Appendix B
IT-Intensive Case of Springer

See Figs. B.1, B.2, B.3, B.4, B.5, B.6 and B.7.

Fig. B.1 Julius Springer, Founder and CEO (left), 1930s (approx.)

Fig. B.2 Early book published by Julius Springer in late 1800s

Fig. B.3 Memorial board in Berlin, Germany

Fig. B.4 Title page of The Science of Nature (an interdisciplinary journal launched by Springer in 1913 to compete with the British periodical Nature)

Fig. B.5 Springer offices in New York (USA, est. 1964) and Chennai (India, est. 1996)

Fig. B.6 Springer logos for different projects (1996–2015)

Fig. B.7 Scanning old books ("Springer book archives" project, 2010)

Appendix C
IT-Intensive Case of IGI Global

See Figs. C.1, C.2 and C.3.

Fig. C.1 IGI Global's
Founder, Dr. Mehdi
Khosrow-Pour, D.B.A

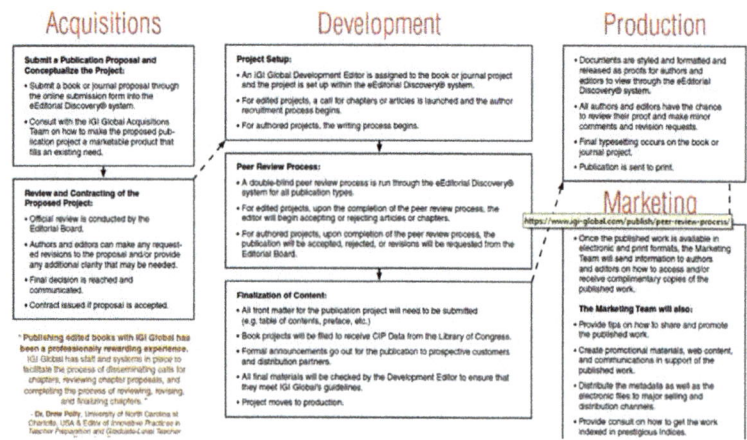

Fig. C.2 IGI Global's publishing process

Fig. C.3 IGI Global's
manuscript publishing
process

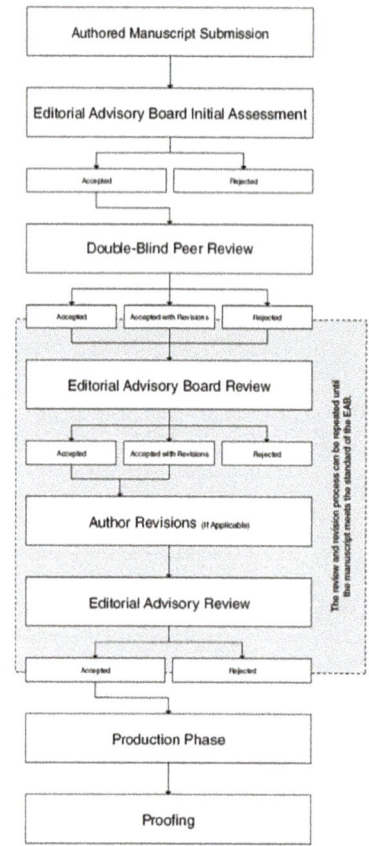

Appendix D
IT-Intensive Case of Dodo Pizza

DodoIS as the Dodo Pizza Company Understands it

Key success ingredients of this multinational delivery-oriented pizza business are: "innovation, deep digital integration and automation" ([http://www.dodofranchise.com/]).

The core of this business is cloud-based software, which coordinates processes so that they are streamlined, and optimizes resources. This makes the business transparent and competitive.

DodoIS is a tightly integrated set of cloud-based applications that control kitchen and delivery activities, collect and process data, and report in real time. This set includes the applications for.

– Production management.
– Assembly line for pizza makers.
– Delivery management (incl. POS).
– Hiring, scheduling, and payroll.
– Inventory management.
– Marketing (incl. promotions, menu boards, etc.).

DodoIS GUI Screenshots

See Figs. D.1, D.2, D.3, D.4, D.5, D.6, D.7, D.8, D.9, D.10, D.11, D.12 and D.13.

DodoIS as a Cloud-Based System

The key advantages of the solution based on Microsoft Azure cloud platform:

© The Editor(s) (if applicable) and The Author(s), under exclusive license to Springer Nature Singapore Pte Ltd. 2022
S. V. Zykov, *IT Crisisology Casebook*, Smart Innovation, Systems and Technologies 300, https://doi.org/10.1007/978-981-19-2231-2

Fig. D.1 Production management in the kitchen (orders immediately appear in the kitchen tablets)

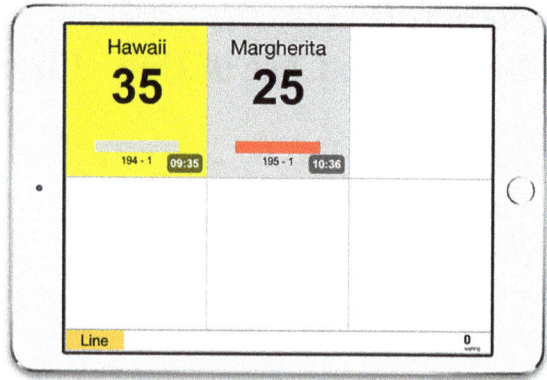

Fig. D.2 Pizza maker's timer (sets priorities and displays preparation time required/left)

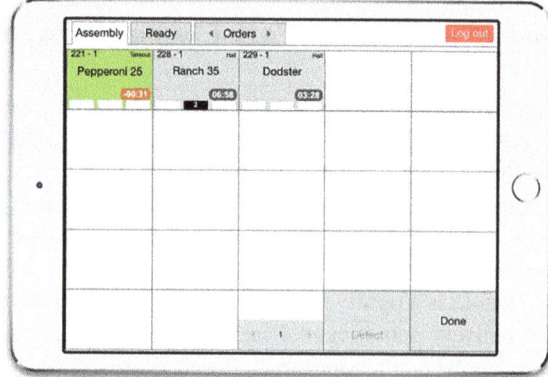

Fig. D.3 Packer's workplace (assists cashiers in order inspection, approval, and dispatching)

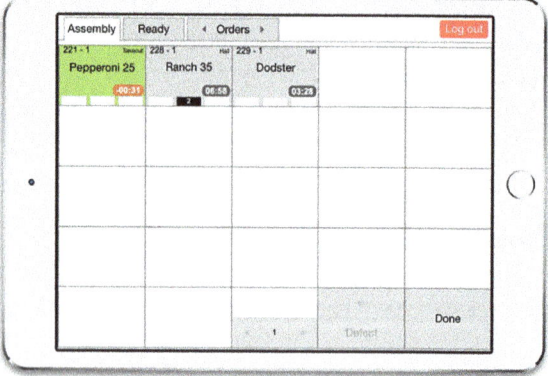

Fig. D.4 Delivery POS
(produces receipt and
packing list, monitors order
status, and delivery time)

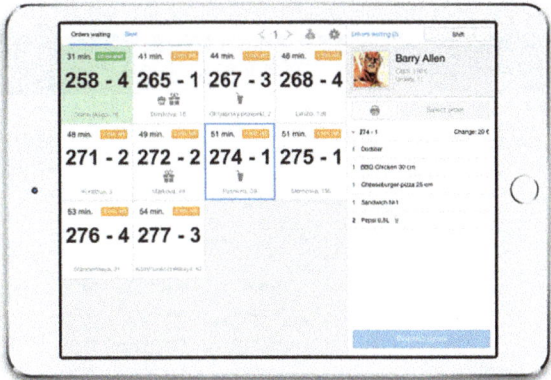

Fig. D.5
Hiring/performance manager
(monitors applicant status
and employee performance)

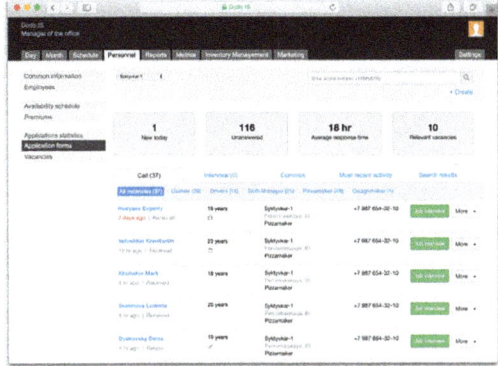

Fig. D.6
Availability/schedule
manager (monitors available
hours, scheduling shifts)

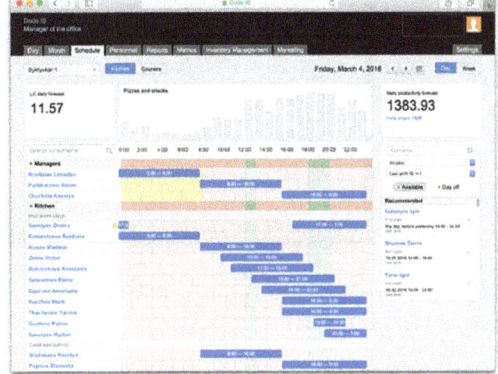

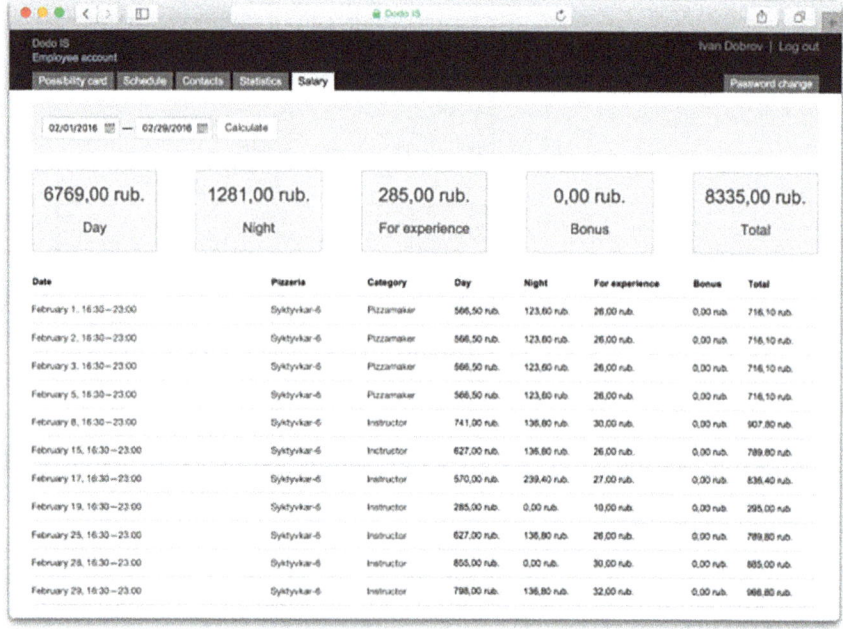

Fig. D.7 Shift manager (assigns shifts, solves scheduling conflicts)

Fig. D.8 Preparation speed manager (displays pizzas made on time and delayed; uses game-like interface for motivation)

- Seamless support of heterogeneous software (operation systems such as iOS, Android, and Windows; applications, languages such as JavaScript, Python, .NET, PHP, Java, Node.js, platforms, tools, and databases) and hardware (tablets, smartphones).
- Free use of open-source software applications.

Fig. D.9 Delivery speed manager (displays delivery time and delays; uses game-like interface for motivation)

Fig. D.10 Inventory manager (ingredients in stock)

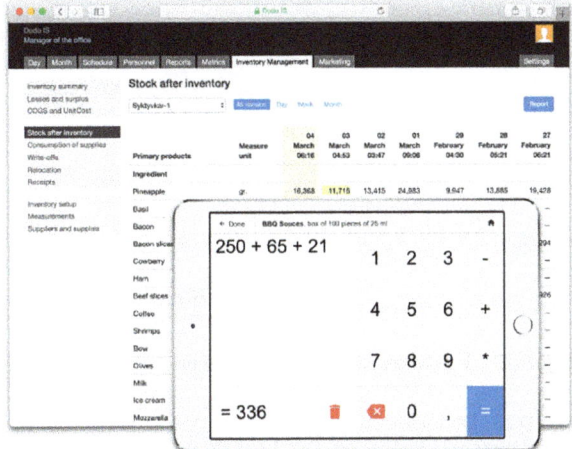

Fig. D.11 Consumption manager (tracks write-offs and their reasons)

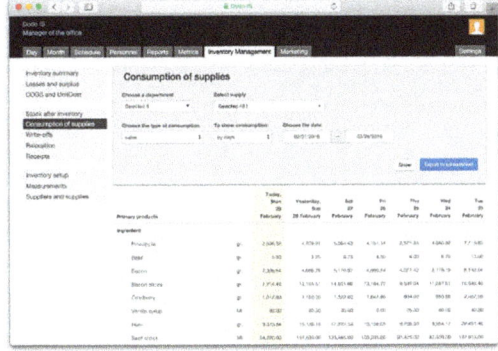

(a)

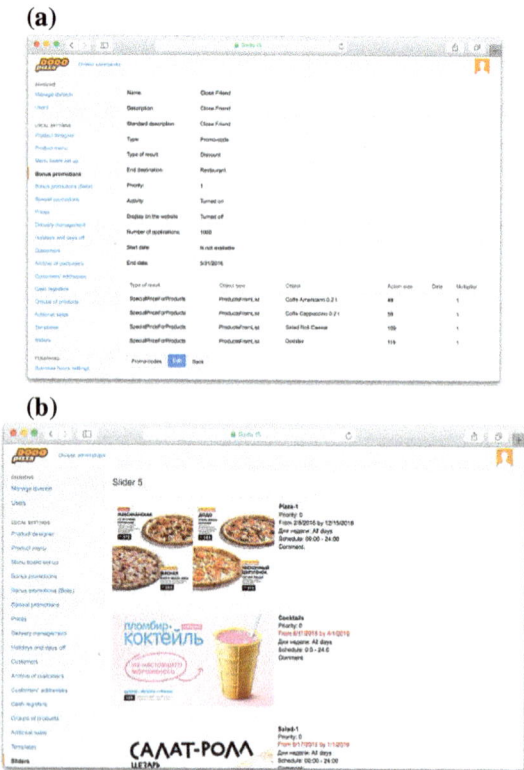

(b)

Fig. D.12 **A, B** Marketing manager (displays and tracks promotions/special offers)

Fig. D.13 DodoIS report
(pizzas prepared by hours;
March 7, 2013;
Pervomaiskaya St.,
Syktyvkar; incl.
promotions/special offers)

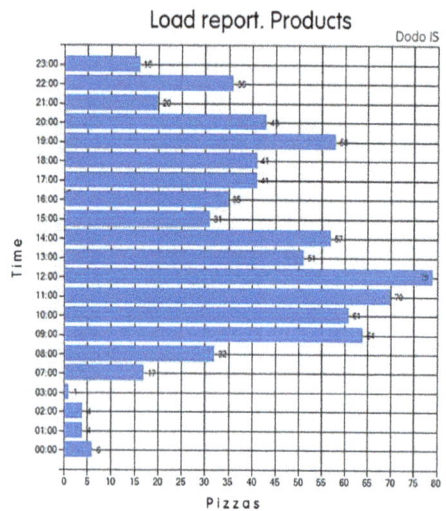

– Uniform service design for a wide range of hardware devices.

Implementing Azure cloud-based DodoIS simplified the application development and packaging processes so that the software can be easily deployed anywhere. Servers are now configured automatically; the applications no longer require multi-server manual installation or complex configuration procedures.

In 2016, Dodo Pizza expands geographically: they opened restaurants in the USA and in China. As such, system administrators easily deployed the Dodo IS providing the same level of system operation stability as in Russia. This happened thanks to the new Azure Container Service by Microsoft that provides scalable clusters of host machines for automated deployment and management of the containerized applications. In other words, there was no need for manual configuring and deploying of the Dodo IS on any of the remote local servers (whether located in USA or China), and therefore the software management costs were minimized.

Dodo Pizza benefits from a strategic synergy of cutting-edge technologies and efficient offline pizza delivery. The technological part helps users to efficiently manage restaurants in a scalable way, as it collects and monitors the key data to analyse the productivity. Due to the new order tracking software system, which is a key element of the DodoIS, all pizzas are prepared from equally fresh ingredients and delivered hot to the customer.

Due to DodoIS, a number of competitive advantages are available for the company; these include the following:

– Direct customer's order propagation from the website to the kitchen tablet PCs.
– Real-time video surveillance of the ordered pizza preparation for each individual customer.

The IT system reduces human factor-based risks, and thereby supports efficient and continuous operation, moving the company from local Russian market to the international food industry.

The Dodo Pizza's KPI and team

See Figs. D.14, D.15, D.16, D.17 and D.18.

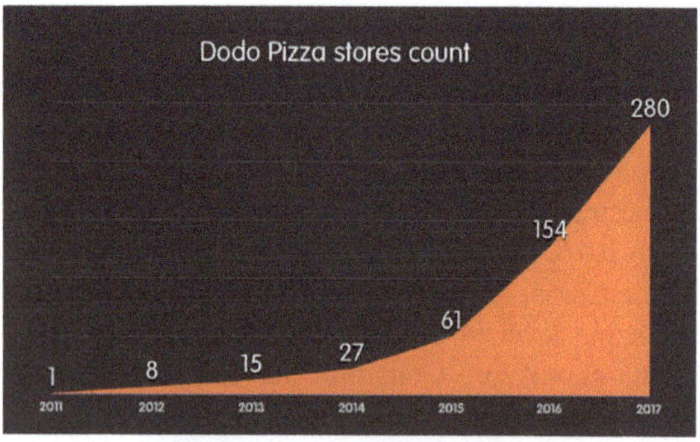

Fig. D.14 Dodo's number
of stores growth
(2011–2017; *Source* 2017
annual report)

Fig. D.15 Dodo's CEO at
Papa John's (which recipe to
follow? The screen at the end
of the line tells that)

Fig. D.16 The motivation
screen of the DodoIS
tracking system

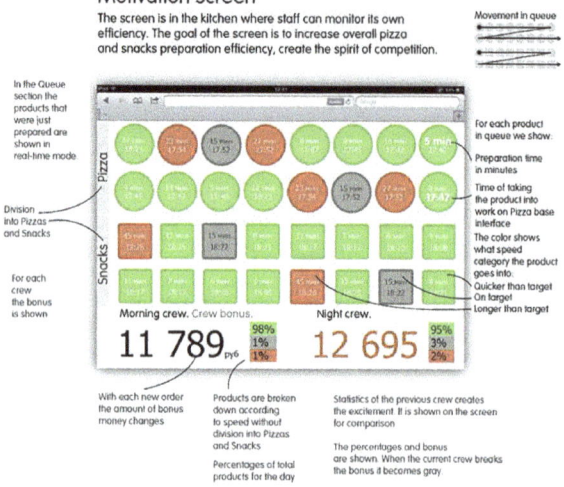

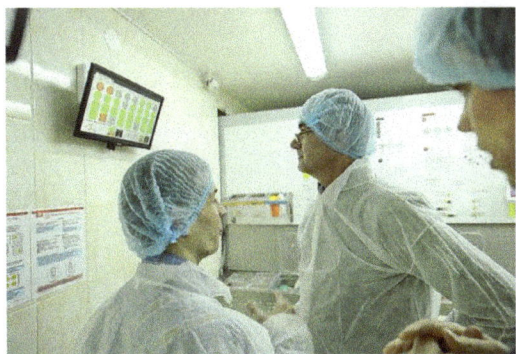

Fig. D.17 Dodo's managers inspecting the motivation screen of the DodoIS tracking system

Fig. D.18 Dodo's kitchen team (*Source* 2017 annual report)

Appendix E
Microsoft Transformation Case

See Figs. E.1, E.2, E.3, E.4, E.5, E.6 and E.7.

Fig. E.1 Satya Nadella, the current Microsoft CEO

Fig. E.2 Steve Ballmer, previous Microsoft CEO

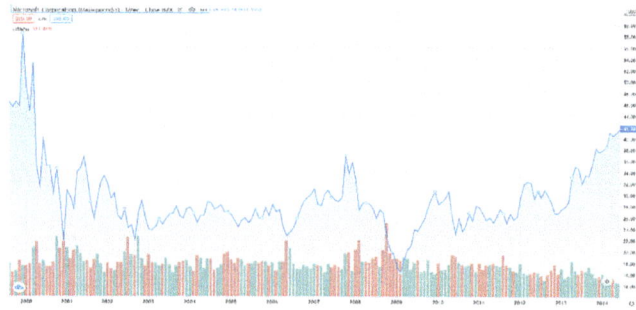

Fig. E.3 Microsoft stock dynamics (2000–2014)

Fig. E.4 Two famous shots. *Left:* Bill Gates presents the Microsoft Tablet PC in 2003, a device he himself dreamed of for many years. *Right:* Steve Jobs, who finalized and perfected the idea of Gates' engineers, at the presentation of the first iPad in 2010

Fig. E.5 Microsoft Surface Neo, a dual-screen notebook tablet. To control the two screens, Windows 10 was upgraded and a new version of Windows 10X was released

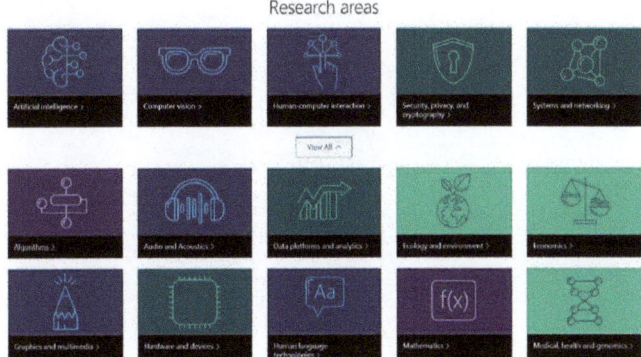

Fig. E.6 Microsoft research areas

Fig. E.7 The underwater data centre before the dive, and its main developers

Glossary

Agile methodology Agile software development is an approach to software development under which requirements and solutions evolve through the collaborative effort of self-organizing and cross-functional teams and their customer/end user. It advocates adaptive planning, evolutionary development, early delivery, and continual improvement, and it encourages rapid and flexible response to change.

Backlog A product backlog is a list of the new features, changes to existing features, bug fixes, infrastructure changes, or other activities that a team may deliver in order to achieve a specific outcome.

Burndown chart A burndown chart is a graphical representation of work left to do versus time. The outstanding work (or backlog) is often on the vertical axis, with time along the horizontal.

Business Intelligence (BI) BI acts as a set of all technologies used to collect and investigate data so that it helps the organization in the decision-making process.

Collective code ownership Collective code ownership abandons any notion of individual ownership of modules. The code base is owned by the entire team and anyone may make changes anywhere.

Communities of Practice (CoPs) Communities of Practice (CoPs) are organized groups of people who have a common interest in a specific technical or business domain. They collaborate regularly to share information, improve their skills, and actively work on advancing the genera knowledge of the domain.

Collaborative wall (information radiator) "Information radiator" is the generic term for any of a number of handwritten, drawn, printed, or electronic displays, which a team places in a highly visible location, so that all team members as well as passers-by can see the latest information at a glance.

Culture All definitions, in general, relate to identifying with the shared mindsets, feeling, shared meaning and characteristics, shared socially developed environments, common ways in which innovations are utilized, and commonly experienced events.

Cross-cultural Cross-culture in the business world refers to a company's efforts to interact effectively with professionals from different backgrounds.

© The Editor(s) (if applicable) and The Author(s), under exclusive license to Springer Nature Singapore Pte Ltd. 2022
S. V. Zykov, *IT Crisology Casebook*, Smart Innovation, Systems and Technologies 300, https://doi.org/10.1007/978-981-19-2231-2

Cross-functional teams A cross-functional team is a group of people with different functional expertise working towards a common goal.

Daily build A daily build or nightly build is the practice of completing a software build of the latest version of a program, daily.

Domain knowledge In software engineering domain knowledge is knowledge about the environment in which the target system operates, for example, software agents.

Exhibit An evidence to support case study, for example, a magazine or newspaper clipping, chart, diagram, or photo.

Explicit knowledge Explicit knowledge (also expressive knowledge) is knowledge that can be readily articulated, codified, stored, and accessed.

Framework A framework, or software framework, is a platform for developing software applications. It provides a foundation on which software developers can build programs for a specific platform.

Globalization The process by which businesses or other organizations develop international influence or start operating on an international scale.

Global Software Development Global Software Development (GSD) is "software work undertaken at geographically separated locations across national boundaries in a coordinated fashion involving real time (synchronous) and asynchronous interaction".

Individualism Individualism is a feel that a person gets re-electing to a personal sense of accomplishment from work, having a work that spares adequate time for individual or for the family members, flexibility to use their own method for work.

Intellectual property Intellectual property (IP) refers to creations of the mind, such as inventions, literary and artistic works, designs and symbols, names and images used in commerce Big-data, and artificial intelligence.

Knowledge management Knowledge management (KM) is the process of creating, sharing, using, and managing the knowledge and information of an organization.

Knowledge transfer Knowledge transfer (KT) is the process of transmitting information that has the source and destination actors and specific context, which serves the purpose of applying and using information to create competitive advantages for the company.

Lightweight methodology A lightweight methodology is a software development method that has only a few rules and practices, or only ones that are easy to follow.

Masculine Masculine characterizes gender roles, for example, in some societies, men are to exhibit masculine characteristics and behaviours as self-confident, tough, intense, and concentrated on objective achievement, while women being more humble, delicate, and concerned with the quality of life.

Modularization The design or production of something in separate sections.

Pair programming Pair programming is an agile software development technique in which two programmers work together at one workstation. One, the driver, writes code while the other, the observer or navigator, reviews each line of code as it is typed in.

Planning game A planning game is a meeting attended by both IT and business teams that is focused on choosing stories for a release or iteration.

Process A series of actions or steps taken in order to achieve a particular end.

Refactoring Code refactoring is the process of restructuring existing computer code—changing the factoring—without changing its external behaviour. Refactoring is intended to improve non-functional attributes of the software.
  Uncontrolled signal amplitude increase.

Retrospective meeting Retrospective, a meeting that is held at the end of an iteration in agile development.

Self-organizing team A self-organizing team is one that does not depend on or wait for a manager to assign work. Instead, these teams find their own work and manage the associated responsibilities and timelines.

Shotgun debugging A process of making relatively undirected changes to software in the hope that a bug will be perturbed out of existence.

Sociocultural variable Sociocultural factors are customs, lifestyles, and values that characterize a society or group.

Software development lifecycle The systems development lifecycle (SDLC), also referred to as the application development lifecycle, is a term used in systems engineering, information systems, and software engineering to describe a process for planning, creating, testing, and deploying an information system.

Sprint A sprint is a set period during which specific work has to be completed and made ready for review.

Tacit knowledge Tacit knowledge resides in the human mind, being a personal asset and is difficult to put it in a formal way, classifies as well as communicate.

Throughput Throughput is the rate of production or the rate at which something is processed.

User Story A user story is a very high-level definition of a requirement, containing just enough information so that the developers can produce a reasonable estimate of the effort to implement it.

Walkthrough A complex example that typically requires the facilitator to lead participants through a series of steps.